SpringerBriefs in Energy

More information about this series at http://www.springer.com/series/8903

Roberto Bonfigli • Stefano Squartini

Machine Learning Approaches to Non-Intrusive Load Monitoring

 Springer

Roberto Bonfigli
Department of Information Engineering
Marche Polytechnic University
Via Brecce Bianche 12
Ancona, Italy

Stefano Squartini
Department of Information Engineering
Marche Polytechnic University
Via Brecce Bianche 12
Ancona, Italy

ISSN 2191-5520   ISSN 2191-5539   (electronic)
SpringerBriefs in Energy
ISBN 978-3-030-30781-3   ISBN 978-3-030-30782-0   (eBook)
https://doi.org/10.1007/978-3-030-30782-0

This Springer imprint is published by the registered company Springer Nature Switzerland AG.
The registered company address is: Gewerbestrasse 11, 6330 Cham, Switzerland

Preface

Research on Smart Grids has recently focused on the energy monitoring issue, with the objective to maximize the user consumption awareness in building contexts on the one hand, and to provide a detailed description of customer habits to the utilities on the other. One of the hottest topics in this field is represented by Non-Intrusive Load Monitoring (NILM), which refers to those techniques aimed at decomposing the consumption aggregated data acquired at a single point of measurement into the diverse consumption profiles of appliances operating in the electrical system under study.

This work reports on a state-of-the-art study of the most promising NILM methods, with an overview of the publically available dataset used for this purpose, as well as a list of all the evaluation metrics used in this research field. Of the proposed methods, those based on the Hidden Markov Model (HMM) and the Deep Neural Network (DNN) have been shown to be the best performing and most interesting from the future improvement point of view. In this work, one method for each category has been selected and the performance improvement achieved is described.

In the HMM based approaches, the Additive Factorial Approximate MAP (AFAMAP) algorithm has shown outstanding capabilities and, therefore, it is nowadays regarded as a reference model. In this work, the AFAMAP algorithm has been extended, by means of a differential forward model, thus complementing the existing differential backward model. Furthermore, an aggregated data examination method has been employed, with the aim of detecting inadmissible working state combinations of appliances, as well as the constraints setting based on the reactive power disaggregation feedback. In a second step, an alternative formulation of the same algorithm is presented, in order to deal with Additive Factorial Hidden Markov Models (FHMM) framework based on bivariate HMM whose emitted symbols are the joint active-reactive power signals. The experiments were conducted on the AMPds dataset, in noised and denoised conditions. Additionally, a user-aided footprint extraction procedure is presented as a facilitated procedure in order to obtain a clean footprint from the aggregated power signal in a real scenario.

In the DNN based approaches, the Denoising Autoencoder (dAE) represents one of the best performing approaches. In this work, this method is extended and improved by conducting a detailed study on the topology of the network, and by intelligently recombining the disaggregated output with a median filter. An exhaustive comparative evaluation is conducted with respect to the AFAMAP algorithm. The experiments have been conducted on the AMPds, UK-DALE, and REDD datasets in seen and unseen scenarios both in presence and in absence of noise. Furthermore, the same method is explored when the input size is increased, including the reactive power component near the active power consumption.

Ancona, Italy   Roberto Bonfigli
Ancona, Italy   Stefano Squartini
June 2019

Contents

Chapter 1
Introduction

Abstract In the recent years, the public awareness on energy saving themes has been constantly increasing. Indeed, the consequences of global warming are now tangible and studies have demonstrated that they are directly related to human activities and their inefficient use of energy and natural resources. The response of governments and public institutions to counteract this trend is to promote policies for reducing energy waste and intelligently use natural resources. The electricity grid is a key component in this scenario: the original electromechanical grid, where the information flow was one-directional, is transforming into the new digital *smart grid* where the information flows from the energy provider to distributed sensors and generator stations and vice versa. Part of this change involves the integration of smart meters in the grid in order to provide detailed consumption information both to the consumers and to the energy provider.

Keywords Energy awareness · Smart grids · Consumer · Energy provider · Non-intrusive load monitoring

In the recent years, the public awareness on energy saving themes has been constantly increasing. Indeed, the consequences of global warming are now tangible and studies have demonstrated that they are directly related to human activities and their inefficient use of energy and natural resources [1–3]. The response of governments and public institutions to counteract this trend is to promote policies for reducing energy waste and intelligently use natural resources. The electricity grid is a key component in this scenario: the original electromechanical grid, where the information flow was one-directional, is transforming into the new digital *smart grid* [4] where the information flows from the energy provider to distributed sensors and generator stations and vice versa. Part of this change involves the integration of smart meters in the grid in order to provide detailed consumption information both to the consumers and to the energy provider.

Indeed, recent studies demonstrated that this fine-grained information is able to provide significant energy savings [5]. On the consumers side, the knowledge of the energy consumption of individual appliances establishes a virtuous behaviour towards a wiser use of electric energy [6, 7]. Studies showed that this can lead to savings greater than 12% with specific appliance feedback and personalized recommendations [5, 8–11]. On the energy provider side, fine-grained information enables the prediction of the power demand, the application of management policies and the prevention of overloading or blackouts over the energy network [12].

Providing detailed consumption information without installing several dedicated meters requires intelligent methods able to infer the energy consumed by individual appliances with minimal metering points. Non-intrusive load monitoring (NILM) denotes the class of methods and algorithms able to perform this task by using the electrical parameters measured in a single-point [5, 13, 14]. Originally developed in the seminal work by Hart [15], NILM has been an active area of research in the last years. The most promising approaches recently presented in the literature are based on machine learning algorithms, and their general scheme consists in extracting significant features from the measured electrical parameters and then estimating the appliance specific active power signal by using a supervised or unsupervised algorithm [13, 16].

As aforementioned, machine learning techniques have become a popular choice for NILM, since they showed significant disaggregation performance: in particular hidden Markov models (HMMs) [17–29] and Neural Networks (NN) [30–34], despite other approaches as graph-based signal processing [35], Support Vector Machines (SVM) [36], k-Nearest Neighbours [36] and Decision Trees [37] have been successfully employed for NILM. This book is focused on the first two categories.

The majority of the approaches employ the active power (P_a) consumption, but other signals can be also effectively used, in order to have a better representation of the electric load, such as reactive power (P_r). This book is focused on the exploitation and the integration of the reactive component of the power consumption within the approaches under study, in order to improve their performance.

The outline of the book is the following. In Chap. 2, the NILM is introduced, with an update state of the art of the approaches in literature and the dataset publicly available for the experiments. Chapter 3 describes the fundamental notion on the hidden Markov Model and neural network paradigm, entering in details for the models parameters meaning and the training algorithm for their estimation. The details of the proposed disaggregation algorithm are presented in Chaps. 4 and 5, respectively, AFAMAP [21] and the denoising Auto Encoder [31]. In both chapters, the improvement of the method and the experimental setup are described, with a discussion on the related results. For both the approaches, the integration of the reactive power component has been proposed. Finally, Chap. 6 concludes this book and presents future developments.

Chapter 2
Non-intrusive Load Monitoring

Abstract The issues relating to the energy conservation and efficiency have gained a role of great importance, from the point of view of both the consumer and the energy provider. Furthermore, over the years, the infrastructures for energy distribution have undergone an ageing process, which have led to the study of the possibility in smart grids implementation, in which a set of information from detection and network management systems can be transmitted in addition to energy.

Useful information, about the characteristics and operating behaviour of an electrical system, can be obtained by means of the power consumption analysis, in order to predict the power demand (load forecasting), to apply management policies and to avoid overloading or blackouts over the energy network. Similarly, from the user perspective, the lifestyle of the people in a house can be predicted by the energy consumption analysis, allowing to implement policies for advantageous time tariffs.

Over the years, several studies have demonstrated that the energy consumption awareness (i.e., which appliances are operating at a certain time instant and how much electrical power they are consuming) influences the user behaviour. Specifically, the awareness conducts to moderate energy consumption, resulting in monetary savings and reduction of the energy required to the provider. Furthermore, applying this consideration to commercial or industrial environments, it may provide larger energy saving.

In the struggle to improve the energy efficiency of residential environments, the availability of information about the appliances in use can support automated optimization approaches.

Load monitoring has become a challenging problem, and several techniques have been studied to solve it. This work is focused on Non-Intrusive Load Monitoring (NILM) algorithms, which aim to separate the aggregated energy consumption signal, measured in a single centralized point, in the individual signals from each appliance, using a simple hardware but smart software algorithms. This solution replaces a distributed smart socket grid inside the house, resulting in lower implementation costs and less invasive solutions for the end user.

© The Author(s), under exclusive license to Springer Nature Switzerland AG 2020

R. Bonfigli, S. Squartini, *Machine Learning Approaches*
to Non-Intrusive Load Monitoring, SpringerBriefs in Energy,
https://doi.org/10.1007/978-3-030-30782-0_2

Keywords Non-intrusive load monitoring · State of the art · Hidden Markov model · Deep neural network · Energy dataset · Evaluation metric

2.1   Problem Statement

The NILM problem can be formulated as follows: let $\overline{y}(t)$ be the aggregated signal measured at the time index t. Without lack of generality, here it is supposed that $\overline{y}(t)$ represents the active power. $\overline{y}(t)$ can be expressed as the sum of the active power contributions of each appliance:

$$\overline{y}(t) = \sum_{i=1}^{N} y^{(i)}(t) + e(t), \tag{2.1}$$

where N is the number of appliances, $y^{(i)}(t)$ is the individual contribution of appliance i and $e(t)$ is a noise term. The NILM problem is, thus, the task of finding the individual appliance contributions $y^{(i)}(t)$ given only the aggregated measurement $\overline{y}(t)$. In a *denoised* scenario [27], the term $e(t)$ is zero, while in a *noised* scenario $e(t)$ can comprise both measurement noise and the contributions of other appliances (e.g., unknown or always-on appliances). The noise term can be treated as a single additional appliance or as an actual noise contribution.

The NILM is classified as a *Blind Source Separation* (BSS) problem. Specifically, it is categorized as a single-channel overcomplete BSS, since the signals, i.e. the power consumptions, flow through the electric line from the multiple loads to the unique sensor, i.e. the smart meter. In the case of analysing the active power consumption, the meter samples the aggregate current flowing in the electric line, and multiplying it with the voltage values, which is approximately a fixed value, it allows to calculate the aggregate power consumption. In order to exploit the reactive power data, both current and voltage have to be sampled in the electric line, in order to recover the phase between them, which is the crucial information for the reactive power calculation. The introduction of the second meter allows to reach a higher level of representation of the problem, which reverse in a more accurate disaggregation results.

2.2   State of the Art

This section presents an overview of the recent literature on NILM.

Several approaches are proposed in the literature, which could be gathered in two main categories, as discussed in [44] (Fig. 2.1): *load classification* and *source separation*. In the former, the disaggregation is achieved by a first step of signature detection, which corresponds to the activation of a specific appliance,

Fig. 2.1 NILM paradigm: the overall power load, given as input, is disaggregated in output signals, each one representing an appliance contribution (i.e., dishwasher, microwave and washing machine)

and a second step of event classification by means of appliance model, previously trained over some training data. In the latter approach, the disaggregation is achieved by recovering the source signals, which in this case correspond to the electrical consumption of each appliance in the network.

For the load disaggregation purpose, defining how the specific appliance in the circuit can be identified within the aggregated signal is fundamental: for this reason a *signature* is defined as a particular trait over the aggregated signal that can be associated to a specific appliance, which can be exploited to permit the disaggregation goal.

For different application, signature is defined in different ways. Two main signature categories can be found in the literature: *steady-state* and *transient* signature. The former [15, 17, 18, 21, 22, 24–26, 31–33, 35] relates to changing operation state of the appliance (i.e., when an appliance is turned on/off), which is reflected on the power characteristics: the value of power measurement is stable in time until the appliance changes operation state, thus this kind of signature can be captured with low frequency sampling (respectively in the order of Hz). Neverthless, if rapid state changes occurs, the low resolution may result unsuitable. The latter [13, 16, 34, 45–50] is based on the transient phenomena between steady states: high frequency noise in electrical current or voltage, as a result of an appliance changing operation state, can be exploited to recognize the different appliances. For this purpose, a high sampling rate is required (respectively in the order of kHz), with a more complex and costly hardware equipment [16]. This explains why the scientific community devoted particular attention to steady-state approaches.

The necessity of the user intervention for creating appliance models distinguishes supervised from unsupervised approaches [51]. The first implies the availability of the individual signals of each appliance. In a real operating scenario, this translates into requiring support by the user, that should sequentially switch on the appliance of interest and switch off the remaining [15]. In this book, this requirement has been partially reduced by allowing selected appliances (e.g., the fridge) to remain operational while signatures of the other appliances are being created, as described in Sect. 4.4. The latter have been the preferred choice in the literature, since they represent the most convenient approach for end-users. Unsupervised techniques provide the means to automate the learning process, thus being completely transparent to the user. Furthermore, they are capable of dynamically adapting to the power system changes over time (i.e., addition, removal, or substitution of appliance) [52]. However, their major shortcoming is represented by the inability to apply an appropriate label to the disaggregated signals. Different approaches try to overcome these limitations by exploiting the information contained on a generic labelled dataset and generalizing to unseen household data by using an unsupervised algorithm [18].

A comparison between the steps required for a supervised and an unsupervised approach is depicted in Fig. 2.2: in order to achieve the load disaggregation purpose, for the former approach individual appliance data are necessary to create models used by the NILM algorithm, while for the latter approach no information other than the aggregate data is required. Although various techniques have been already presented in the literature, which obtain reasonable performance, most of them are based on supervised algorithms (i.e., require individual appliance data for model training, prior to the system deployment), thus their functioning depends on the user intervention and the a-priori knowledge of the power system parameters in which they are working. In order to prevent these inconveniences, unsupervised NILM techniques have been developed: these approaches do not require individual appliance data and the models information is captured only using the aggregated load, without the user intervention. Furthermore, the unsupervised approaches are independent from the number of the appliances forming the aggregated load and

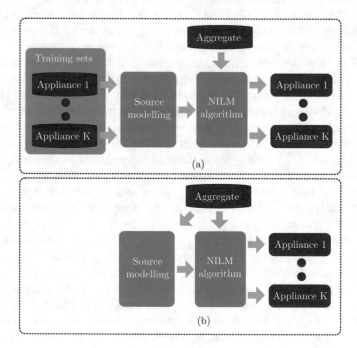

Fig. 2.2 Comparison of supervised (**a**) and unsupervised (**b**) method

capable of dynamically adapt to the power system changes over time (i.e., addition, removal or substitution of appliance).

For a recent review and a taxonomy, please refer to [13, 16, 53, 54]. In [55] different approaches are described, also with an overview over metering equipment for data logging.

Some techniques are publicly available implemented within the NILMTK toolkit [56] and the NILM Eval framework [57].

Among unsupervised approaches, the ones based on FHMMs have been devoted particular attention in the last years. One of the earliest works on the topic has been presented in [17] by Kim and colleagues. The key idea is to model each appliance with independent parallel HMM each contributing to the aggregate power. The framework is assessed by using the steady-state real power signal, but it allows multidimensional features as input. In [18], the authors employ HMMs in a Bayesian framework in order to combine multiple models and form a general model of an appliance. Labelled data are required in the training phase and then appliance specific models are tuned on aggregate data without requiring user intervention. In the literature, particular attention has been devoted to the algorithm proposed by Kolter and Jaakkola [21], since it showed noteworthy performance with a reasonable computational complexity. The Additive Factorial Approximate Maximum a Posteriori (AFAMAP) algorithm is an efficient method, based on an optimization problem, for the inference of the working states combination in the Factorial Hidden Markov Model framework. The authors introduced the AFAMAP algorithm, where they constrain the posterior probability to require only one HMM change state at any given time. Semi-Markov models are combined with Hierarchical Dirichlet Process in [28] for inferring both the state complexity of the models and the duration of the distributions. The authors use the active power as input feature and evaluate the performance on the five most consuming appliances of the REDD dataset [29]. Makonin and colleagues in [27] proposed the sparse Viterbi algorithm for disaggregating the active power online and in real-time. Sparse Viterbi exploits the matrix sparsity in HMMs and it was evaluated on the AMPds [58] and REDD [29] datasets. Aiad and Lee [51] augmented FHMMs with additional chains for modelling possible interactions among the appliances. The algorithm operates on the active power input feature and it was evaluated on the REDD dataset. The work in [22] introduces an FHMM model with unbounded number of chains, and states for each chain as well. In [23] the authors introduce Hierarchical FHMM with the aim of overcoming the device independence assumption and the one-at-time condition. The algorithm operates on the steady-state active power signal by clustering the signals of correlated devices and then by training HMM models on the identified clusters (denoted as "super devices"). In the disaggregation phase, inference is performed with AFAMAP on the super devices, and the result is mapped back to the original device by using the state relation table learned during the training phase. Compared to the original AFAMAP algorithm on the REDD and Pecan datasets, the method proposed by the authors provides significant performance improvements. Zhong et al. [24] incorporate domain knowledge in the FHMM in the form of signal aggregate constraint. In the NILM scenario, this translates into constraining the total energy consumed in a day by an appliance to be close to a predefined value. The algorithm was assessed on the Household Electricity Survey dataset and compared to the Additive Factorial HMM and the AFAMAP algorithms. The results showed that the method indeed achieves better performance in terms of disaggregation error. In a different work [25], the same authors introduce interleaved factorial non-homogeneous hidden Markov model (IFNHMM), where the transition probabilities of the models are supposed time variant in order to represent the different pattern of usage of an appliance during the day. In addition, at each time step only one chain is allowed to change. The algorithm presented in [26] combines FHMM and Subsequence Dynamic Time Warping (SDTW). The FHMM is employed in the first stage to identify only the ON and OFF state of each appliance. SDTW, then, is applied iteratively to extract the final output. The authors propose both a supervised and semi-supervised version of the algorithm, with the latter employing the aggregate signal and consumption diaries to extract the appliance signatures.

The works presented above perform load disaggregation by using the active power as the only input feature. Differently, in [20], the authors propose a structural variational approximation method and they evaluated the combination of five features: active and reactive power, power factor, and the active and reactive power standard deviation calculated in a window of five samples. The algorithm is evaluated in a "denoised scenario", for different combinations of low-power appliances (e.g., laptop, desk lamp, LCD monitor). Instead of using only electrical parameters, in [59] the authors proposed the inclusion of contextual information represented by the timing-usage statistics and the presence of the user in the house. The disaggregation algorithm is based on AFAMAP and Conditional FHMMs, and the experiments are conducted on the Tracebase dataset augmented with synthetic contextual information.

Among the techniques appeared in the literature, Deep Neural Networks (DNN) have been devoting particular attention in the last years, since they exhibited noteworthy performance for load disaggregation [31–33]. In [32], the authors proposed an approach based on Long Short-Term Memory (LSTM) neural networks [60]. The algorithm consists in training a neural network for each appliance in order to predict a sample of the disaggregated active power from a segment of aggregated data. Neural networks have been combined with HMMs in [33]: the emission probabilities of the HMM are modelled by a Gaussian distribution for state representing the single load, and by a DNN for state representing the aggregated signal. Similarly to [32], LSTMs have been also employed in [31], this time combined with convolutional layers at the input of the network to extract the features of the signal directly from raw data. In the same paper, NILM is treated also as a noise reduction problem, where the clean signal is represented by the disaggregated appliance profile, and the noise signal by the remaining profiles and the measure-ment noise. Noise reduction is performed by using a denoising autoencoder (dAE) composed of convolutional and fully connected layers that estimates the appliance profile from the aggregated noisy signal. An additional approach proposed in [31] uses a neural network that estimates the start time, the end time and the mean power demand of each appliance. In the experiments conducted by the authors on the UK recording Domestic Appliance-Level Electricity dataset (UK-DALE) [61], they demonstrated that the most performing approach is represented by the dAE network, that outperformed both the other DNN architectures, and the FHMM method proposed in [29].

A different approach has been proposed in [62], where the algorithm employs motif mining to identify recurring events. In particular, based on the a-priori knowledge of the number of devices, it operates by firstly removing the appliances that are always on. Then, it identifies the steady-states power levels with a Dirichlet process Gaussian Mixture Model, and it detects repetitive sequences of power level changes. The probabilistic sequential mining stage discovers devices with several sequential power levels. The algorithm operates by firstly clustering power levels according to the time of day and day of the week. Finally, the motif mining stage finds repetitive episodes in the time series. On average, the results obtained on the REDD dataset showed a superior performance with respect to the AFAMAP algorithm [21]. In [35], the authors propose a graph signal processing (GSP) approach that does not require training data. The GSP paradigm is employed for event detection, clustering and feature matching.

Although in the majority of the approaches the active power (P_a) consumption is employed, other signals can be also effectively used, in order to have a better representation of the electric load, such as reactive power (P_r) consumption, current (I) and voltage (V) signal. Besides using the raw signal, better performance can be achieved introducing a feature extraction stage, in order to represent information at a higher level: different kind of ensemble averages (i.e., mean, variance) or the application of transform operator (i.e., Fourier, Wavelet, ST, Hilbert) are the main features employed. In addition, other quantities can be extracted to represent specific information about the appliance usage, such as cycling frequency and temporal duration usage, or indicator representative of the appliance electric circuit, such as current/voltage harmonic distortion.

2.3 Datasets

Every problem to be solved with machine learning and data mining techniques requires the availability of data for algorithm parametrization: the ability to access public dataset, representative of a real scenario, allows to test the approaches, in order to evaluate the effective benefit in real applications, and to compare the performance of existing approaches on a common comparison basis. In order to evaluate the effectiveness of the algorithms and the performance about the disaggregation task, both aggregate and appliance specific data, which represent the ground truth, are required.

Comparison between the datasets, highlighting their main characteristics, such as duration, number of houses and signal sampling frequency is shown in Table 2.1.

Table 2.1 Comparison of household energy datasets

Contribution	Dataset	Location	Duration per house	Number of houses	Appliance sample resolution	Aggregate sample resolution
[29]	REDD	USA	3–19 days	6	3 s	1 s and 15 kHz
[63]	BLUED	USA	8 days	1	Transition label	12 kHz
[64]	UMass Smart	USA	3 months	3	1 s	1 s
[65]	Tracebase	DE	N/A	15	1–10 s	N/A
[66]	Pecan Street	USA	7 days	10	1 min	1 min
[67]	HES	UK	1 or 12 months	251	2 or 10 min	2 or 10 min
[58]	AMPds	CDN	1 year	1	1 min	1 min
[68]	iAWE	IND	73 days	1	1 or 6 s	1 s
[61]	UK-DALE	UK	3–17 months	4	6 s	1–6 s and 16 kHz
[69]	GreenD	AT/IT	1 year	9	1 s	1 s
[70]	COMBED	IND	18 months	8	30 s	30 s
[57]	ECO	CH	8 months	6	1 s	1 s
[71]	BERDS	USA	1 year	N/A	20 s	20 s
[72]	SustData	PT	5 years	50	50 Hz	50 Hz

This comparative table is an extension of the proposed one in [56], with an update considering the recent datasets published in the last year.

From a geographic point of view, in most cases the datasets are recorded in the USA, with some examples for European countries (i.e., Germany, United Kingdom, Austria, Italy, Switzerland and Portugal), besides Canada and India. The recording coming from different country could lead to mismatching between electric quantity (i.e., the RMS voltage value is 220 V in Europe and 110 V in the USA), thus attention needs to be paid when different datasets are used in the same system development. It can be noticed that the consumption recordings last several days or few months for many contributions. Nevertheless, several datasets contain recordings one or more years long: in these cases it is possible to study the human behaviour over a long time, comprising the effect of seasonal changes on consumption. In addition, only in [67, 72] a high number of houses is present, which lead to studies about power circuit behaviour in different households. Regarding the sampling frequency of the aggregate data and specific appliance signals, there is a common trend about using a sampling interval between 1 s and 1 min. Only in [29, 63, 73] a higher sampling frequency, in order of kHz, is used, which allows the development of more sophisticated algorithm: the availability of a higher data resolution allows to examine transient phenomena, which can be used for a more complete description of the problem.

In [56] an open source toolkit is presented, called *NILMTK*, useful to evaluate NILM algorithms in a simple way over different datasets. The toolkit contains a data importer for each dataset, a set of preprocessing and statistics functions, a list of some disaggregation algorithms and a set of metrics to evaluate the performance of such algorithms. The complete processing pipeline is reported in Fig. 2.3.

2.4   Evaluation Metrics

The metrics chosen for the performance evaluation have to represent both the aspects of the disaggregation problem: the classification of the switching activity of the appliances and the accuracy of the disaggregated profiles compared to the ground truth appliance consumption [73].

In order to evaluate both aspects of the NILM problem, algorithms have been evaluated by using the following metrics:

- Energy-based Precision ($P^{(E)}$), Recall ($R^{(E)}$), and F_1-Measure ($F_1^{(E)}$) [21];
- Normalized Disaggregation Error (NDE) [21];
- Normalized Error in Assigned Power (NEP) [73];
- State-based Precision ($P^{(S)}$), Recall ($R^{(S)}$), F_1-Measure ($F_1^{(S)}$);
- Matthews Correlation Coefficient (MCC) [73, 74].

Energy-based Recall measures the part of the power consumption that has been correctly classified, whereas the Precision measures the amount of power assigned

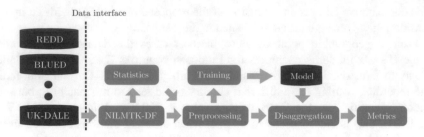

Fig. 2.3 The processing pipeline of NILMTK. Courtesy of Batra et al.

to an appliance that actually belongs to it. Considering the i-th appliance, $P_i^{(E)}$ and $R_i^{(E)}$ are calculated as follows:

$$P_i^{(E)} = \frac{\sum_{t=1}^{T} \min\left(\hat{y}^{(i)}(t), y^{(i)}(t)\right)}{\sum_{t=1}^{T} \hat{y}^{(i)}(t)}, \quad R_i^{(E)} = \frac{\sum_{t=1}^{T} \min\left(\hat{y}^{(i)}(t), y^{(i)}(t)\right)}{\sum_{t=1}^{T} y^{(i)}(t)}, \quad (2.2)$$

where $\hat{y}^{(i)}(t)$ is the disaggregated power consumption signal, $y^{(i)}(t)$ is the ground truth appliance power consumption signal and T is the total number of samples. In order to evaluate the total performance of the disaggregation algorithm, the metric average across the appliances is computed as follows:

$$P^{(E)} = \frac{1}{N} \sum_{i=1}^{N} P_i^{(E)}, \quad R^{(E)} = \frac{1}{N} \sum_{i=1}^{N} R_i^{(E)}. \quad (2.3)$$

The F_1-Measure is calculated as the geometric mean between Precision and Recall:

$$F_1^{(E)} = 2 \frac{P^{(E)} R^{(E)}}{P^{(E)} + R^{(E)}}. \quad (2.4)$$

The Normalized Disaggregation Error (NDE) [21] provides a direct measure of the ability of the algorithm of reconstructing the active power profiles, and it is defined as:

$$NDE = \sqrt{\frac{\sum_{t=1}^{T} \sum_{i=1}^{N} \left(y^{(i)}(t) - \hat{y}^{(i)}(t)\right)^2}{\sum_{t=1}^{T} \sum_{i=1}^{N} \left(\hat{y}^{(i)}(t)\right)^2}}. \quad (2.5)$$

The Normalized Error in Assigned Power (NEP) measures the deviation of the estimated power $\hat{y}^{(i)}(t)$ from the true power $y^{(i)}(t)$ normalized by the total energy consumption of the appliance. Considering appliance i, NEP is calculated as follows:

$$\text{NEP}_i = \frac{\sum_{t=1}^{T} |y^{(i)}(t) - \hat{y}^{(i)}(t)|}{\sum_{t=1}^{T} y^{(i)}(t)}. \tag{2.6}$$

State-based metrics are defined based on the actual and predicted state of an appliance. More in details, considering appliance i, true positives (TP), false positives (FP), false negatives (FN) and true negatives (TN) are defined as follows:

$$\text{TP}_i = \sum_{t=1}^{T} \text{AND}(x^{(i)}(t) = \text{on}, \hat{x}^{(i)}(t) = \text{on}), \tag{2.7}$$

$$\text{FP}_i = \sum_{t=1}^{T} \text{AND}(x^{(i)}(t) = \text{off}, \hat{x}^{(i)}(t) = \text{on}), \tag{2.8}$$

$$\text{FN}_i = \sum_{t=1}^{T} \text{AND}(x^{(i)}(t) = \text{on}, \hat{x}^{(i)}(t) = \text{off}), \tag{2.9}$$

$$\text{TN}_i = \sum_{t=1}^{T} \text{AND}(x^{(i)}(t) = \text{off}, \hat{x}^{(i)}(t) = \text{off}), \tag{2.10}$$

where $x^{(i)}(t)$ and $\hat{x}^{(i)}(t)$ are, respectively, the actual and the predicted state of appliance i at the time index t. Appliance i is considered in the "on" state if $y^{(i)}(t)$ exceeds a predefined threshold. Generally, the threshold varies with the appliance and it assumes the same value used for extracting the activations within the ground truth power consumption [31]. State-based Precision and Recall are defined as:

$$P_i^{(S)} = \frac{\text{TP}_i}{\text{TP}_i + \text{FP}_i}, \quad R_i^{(S)} = \frac{\text{TP}_i}{\text{TP}_i + \text{FN}_i}, \tag{2.11}$$

In the case of multi-state models, e.g. HMM or FSM, the *state based* metric considers the ability of the system to infer the exact state of evolution of each HMM in the model: for the i-th appliance, the multiclass confusion matrix is built by comparing, for each time instant $t = 1, 2, \ldots, T$, the disaggregation variables $\boldsymbol{\xi}^{(i)}(t)$ value assumed in the problem solution, with the exact evolution state $x^{(i)}(t)$, defined as the *ground truth*. Each class corresponds to a state $j = 1, \ldots, m_i$ of the i-th HMM. Since that the values in $\boldsymbol{\xi}^{(i)}(t)$ are not-integral, the computed confusion matrix is soft weighted, similar to the fuzzy-logic [75]. For each class, the Precision $P_i^{(j)}$ and Recall $R_i^{(j)}$ are computed, then the average between the classes evaluates the medium performance for each HMM:

$$P_i^{(S)} = \frac{1}{m_i} \sum_{j=1}^{m_i} P_i^{(j)}, \quad R_i^{(S)} = \frac{1}{m_i} \sum_{j=1}^{m_i} R_i^{(j)}. \tag{2.12}$$

Finally, state-based F_1-Measure is given by:

$$F_1^{(S)} = \frac{2 P^{(S)} R^{(S)}}{P^{(S)} + R^{(S)}}, \quad \text{with} \quad P^{(S)} = \frac{1}{N} \sum_{i=1}^{N} P_i^{(S)}, \ R^{(S)} = \frac{1}{N} \sum_{i=1}^{N} R_i^{(S)}. \tag{2.13}$$

The Matthews Correlation Coefficient is defined as:

$$\text{MCC}_i = \frac{\text{TP}_i \text{TN}_i - \text{FP}_i \text{FN}_i}{\sqrt{(\text{TP}_i + \text{FP}_i)(\text{TP}_i + \text{FN}_i)(\text{TN}_i + \text{FP}_i)(\text{TN}_i + \text{FN}_i)}}, \tag{2.14}$$

and

$$\text{MCC} = \frac{1}{N} \sum_{i=1}^{N} \text{MCC}_i. \tag{2.15}$$

MCC assumes values in the range $[-1, 1]$, with $+1$ representing perfect prediction, 0 random prediction and -1 total disagreement between the ground truth and the prediction.

In the case of the metrics are evaluated for a signal window w_f with $f = 1, 2, \ldots, F$, the metrics are averaged over the windows, since the performance is evaluated over the entire dataset:

$$P_i^{\{(S),(E)\}} = \frac{1}{F} \sum_{f=1}^{F} P_i^{\{(S_f),(E_f)\}}, \quad R_i^{\{(S),(E)\}} = \frac{1}{F} \sum_{f=1}^{F} R_i^{\{(S_f),(E_f)\}}. \tag{2.16}$$

2.5 Remarks

Regarding the datasets, the difference among the many appeared in the literature is highlighted, in terms of amount of recorded data, number of houses and sampling frequency. All these parameters, together with the characteristics of the available smart meter providing the aggregate consumption data in the operating scenario, strongly influence the choice of the NILM technique and therefore the dataset for algorithm design and optimization must be carefully selected.

Chapter 3
Background

Abstract The computers are able to perform complex calculus operations in a short amount of time. However computers cannot compete with humans in dealing with: common sense, ability to recognize people, objects, sounds, comprehension of natural language, ability to learn, categorize, generalize.

Therefore, why does the human brain show to be superior w.r.t common computers for these kind of problems? Is there any chance to mimic the mechanisms characterizing the way of working of our brain in order to produce more efficient machines?

In the field of signal analysis, the aim is the characterization of such real-world signals in terms of *signal models*, which can provide the basis for a theoretical description of a signal processing system. They are potentially capable of letting us learn a great deal about the signal source, without having to have the source available.

Therefore, in this chapter two families of modelling technique are described, i.e., the hidden Markov models (HMM) and the Deep Neural Network (DNN). After a theoretical description, the algorithms used for their parameter estimation are described, with a focus on the most widely model structure used in the field of the NILM.

Keywords Machine learning · Hidden Markov Model · Baum Welch algorithm · Deep neural network · Stochastic gradient descent

3.1 Hidden Markov Model (HMM)

Within the multiple technique available, there are several possible choices for what type of signal model is used for characterizing the properties of a given signal. The most widely used categorization gathers the methods in deterministic models and statistical models. In this chapter, it is interesting to explore the statistical models, which try to characterize only the statistical properties of the signal.

R. Bonfigli, S. Squartini, *Machine Learning Approaches to Non-Intrusive Load Monitoring*, SpringerBriefs in Energy, https://doi.org/10.1007/978-3-030-30782-0_3

The underlying assumption of the statistical model is that the signal can be well characterized as a parametric random process, and that the parameters of the stochastic process can be determined (estimated) in a precise, well-defined manner. One type of stochastic signal model is the *hidden Markov model* (HMM). This model is based on some theoretical fundamentals. The treatise followed in this chapter is inspired by Rabiner [76].

Firstly, a *discrete Markov process* need to be introduced. It is a system which may be described at any time as being in one of a set of m distinct states, $S_1, S_2, \ldots, S_m$. The time instants associated with state changes are defined as $t = 1, 2, \ldots$, while the actual state at time t as $x(t)$.

In a discrete, first order, Markov chain, this probabilistic description is truncated to just the current and the predecessor state:

$$Pr[x(t) = S_j | x(t-1) = S_i, x(t-2) = S_k, \ldots] = Pr[x(t) = S_j | x(t-1) = S_i].$$
(3.1)

In those processes, the right-hand side of the equation is independent of time. Additionally, a set of state transition probabilities P_{ij} is defined in the form:

$$P_{ij} = Pr[x(t) = S_j | x(t-1) = S_i] \text{ for } 1 \leq i, j \leq m.$$
(3.2)

The state transition coefficients follow the properties:

$$\sum_{j=1}^{m} P_{ij} = 1 \text{ with } P_{ij} \geq 0$$
(3.3)

since they obey to standard stochastic constraints.

The notation to denote the initial state probabilities is the following:

$$\phi_i = Pr[x(1) = S_i] \text{ for } 1 \leq i \leq m$$
(3.4)

The model defined above is classified as an *observable Markov chain*, since the output of the process is the set of states at each instant of time. The extension to *hidden Markov models* (HMM) introduces the fundamental that the observation is a probabilistic function of the state, i.e., the resulting model (which is called a hidden Markov model) is a doubly embedded stochastic process with an underlying stochastic process that is not observable (it is hidden), but can only be observed through another set of stochastic processes that produce the sequence of observations.

Therefore, the elements which constitute an HMM are the following:

- m, the number of states in the model. The states are interconnected in such a way that any state can be reached from any other state (e.g., an ergodic model). The individual states are denoted as $S = \{\ldots\}$, and the state at time t as $x(t)$.

- s, the number of distinct observation symbols per state, i.e., the discrete alphabet size. The individual symbols are denoted as $U = \{\mu_1, \mu_2, \ldots, \mu_s\}$.
- The state transition probability distribution $P = \{P_{ij}\}$, where:

$$P_{ij} = Pr[x(t+1) = S_j | x(t) = S_i] \text{ for } 1 \leq i, j \leq m \tag{3.5}$$

For the special case where any state can reach any other state in a single step, we have $P_{ij} > 0$ for all i, j. For other types of HMMs, we would have $P_{ij} = 0$ for one or more (i, j) pairs.

- The observation symbol probability distribution in state j, $M = \{M_j(k)\}$, where:

$$M_j(k) = Pr[\mu_k \text{ at } t | x(t) = S_j] \text{ for } 1 \leq j \leq m, 1 \leq k \leq s \tag{3.6}$$

- The initial state distribution $\phi = \{\phi_i\}$, where:

$$\phi_i = Pr[x(1) = S_i] \text{ for } 1 \leq i \leq m \tag{3.7}$$

The HMM can be used as a generator to give an observation sequence:

$$Y = \{y(1), y(2), \ldots, y(T)\} \tag{3.8}$$

where each observation $y(t)$ is one of the symbols from U, and T is the number of observations in the sequence.

A complete specification of an HMM requires the definition of two model parameters (m and s), specification of observation symbols, and of the three probability measures P, M and ϕ

$$\lambda = (P, M, \phi). \tag{3.9}$$

The probability of the observation sequence, Y, given the model λ, i.e., $Pr(Y|\lambda)$ for the state sequence $X = \{x(1), x(2), \ldots, x(T)\}$ is defined as:

$$Pr(Y|X, \lambda) = \prod_{t=1}^{T} Pr(y(t)|x(t), \lambda)$$

$$= M_{x(1)}(y(1)) M_{x(2)}(y(2)) \cdots M_{x(T)}(y(T)) \tag{3.10}$$

in which it is assumed the statistical independence of observations.

The probability of such a state sequence X is defined as:

$$Pr(X|\lambda) = \phi_{x(1)} P_{x(1)x(2)} P_{x(2)x(3)} \cdots P_{x(T-1)x(T)}. \tag{3.11}$$

Therefore, the joint probability of Y and X is:

$$Pr(Y, X|\lambda) = Pr(Y|X, \lambda) Pr(X|\lambda) \tag{3.12}$$

The probability of Y (given the model) is obtained by summing this joint probability over all possible state sequences X:

$$Pr(Y|\lambda) = \sum_{all X} Pr(Y|X, \lambda) \, Pr(X|\lambda)$$

$$= \sum_{x(1),x(2),...,x(T)} \phi_{x(1)} M_{x(1)}(y(1)) P_{x(1)x(2)}$$

$$\times M_{x(2)}(y(2)) \cdots P_{x(T-1)x(T)} M_{x(T)}(y(T)) \qquad (3.13)$$

and an efficient procedure to solve the problem is the *Forward-Backward* procedure. The forward variable $\alpha_t(i)$ is defined as:

$$\alpha_t(i) = Pr(\{y(1), y(2), \cdots, y(t)\}, x(t) = S_i|\lambda) \qquad (3.14)$$

and the probability of the partial observation sequence, $\{y(1), y(2), \cdots, y(t)\}$ (until time t) and state S_i at time t, given the model λ, is solved exploiting $\alpha_t(i)$ inductively, following the procedure:

1. Initialization:

$$\alpha_1(i) = \phi_i M_i(y(1)) \text{ for } 1 \le i \le m. \qquad (3.15)$$

2. Induction:

$$\alpha_{t+1}(j) = \left[\sum_{i=1}^{m} \alpha_t(i) P_{ij}\right] M_j(y(t+1)) \text{ for } 1 \le t \le T-1, 1 \le j \le m.$$

$$(3.16)$$

3. Termination:

$$Pr(Y|\lambda) = \sum_{i=1}^{m} \alpha_T(i). \qquad (3.17)$$

On the other hand, a backward variable $\beta_t(i)$ is defined as:

$$\beta_t(i) = Pr(\{y(t+1), y(t+2), \cdots, y(T)\}|x(t) = S_i, \lambda) \qquad (3.18)$$

and the probability of the partial observation sequence from $t+1$ to the end, given state S_i at time t and the model λ, is solved exploiting the $\beta_t(i)$ inductively, following the procedure:

1. Initialization:

$$\beta_T(i) = 1 \text{ for } 1 \le i \le m. \qquad (3.19)$$

2. Induction:

$$\beta_t(i) = \sum_{i=1}^{m} P_{ij} M_j(y(t+1))\beta_{t+1}(j) \text{ for } t = T-1, T-2, \ldots, 1, 1 \le i \le m$$

$$(3.20)$$

The corresponding state sequence has to be chosen.

Additionally, finding the *optimal* state sequence $X = \{x(1), x(2), \ldots, x(T)\}$ associated with the given observation sequence is defined exploiting several possible optimality criteria.

The variable $\gamma_t(i)$ defines the probability of being in state S_i at time t, given the observation sequence Y, and the model λ:

$$\gamma_t(i) = Pr(x(t) = S_i | Y, \lambda) \tag{3.21}$$

$$= \frac{\alpha_t(i)\beta_t(i)}{Pr(Y|\lambda)} = \frac{\alpha_t(i)\beta_t(i)}{\sum_{i=1}^{m} \alpha_t(i)\beta_t(i)} \tag{3.22}$$

where $\alpha_t(i)$ accounts for the partial observation sequence $\{y(1), y(2), \cdots, y(t)\}$ and state S_i at t, while $\beta_t(i)$ accounts for the remainder of the observation sequence $\{y(t+1), y(t+2), \cdots, y(T)\}$ given state S_i at t. The normalization factor $Pr(Y|\lambda) = \sum_{i=1}^{m} \alpha_t(i)\beta_t(i)$ makes $\sum_{i=1}^{m} \gamma_t(i) = 1$.

The most widely used criterion is to find the *single* best state sequence (path) $X = \{x(1), x(2), \cdots, x(T)\}$ for the given observation sequence $Y = \{y(1), y(2), \cdots, y(T)\}$, i.e., to maximize $Pr(X|Y, \lambda)$ which is equivalent to maximizing $Pr(X, Y|\lambda)$. This criterion is satisfied by the *Viterbi algorithm*. The quantity $\delta_t(i)$ is defined as the best score (highest probability) along a single path, at time t, which accounts for the first t observations and ends in state S_i:

$$\delta_t(i) = \max_{\{x(1), x(2), \cdots, x(t-1)\}} Pr[\{x(1), x(2), \cdots, x(t) = i\}, \{y(1), y(2), \cdots, y(t)\}|\lambda],$$

$$(3.23)$$

and, by induction:

$$\delta_{t+1}(j) = [\max_i \delta_t(i) P_{ij}] M_j(y(t+1)) \tag{3.24}$$

To actually retrieve the state sequence, we need to keep track of the argument which maximized $\delta_{t+1}(j)$, for each t and j, via the array $\psi_t(j)$. The procedure follows the steps:

1. Initialization:

$$\delta_1(i) = \phi_i M_i(y(1)) \text{ for } 1 \le i \le m \tag{3.25}$$

$$\psi_1(i) = 0 \text{ for } 1 \le i \le m \tag{3.26}$$

2. Recursion:

$$\delta_t(j) = \max_{1 \leq i \leq m} [\delta_{t-1}(i)P_{ij}]M_j(y(t)) \text{ for } 2 \leq t \leq T, 1 \leq j \leq m \qquad (3.27)$$

$$\psi_t(j) = \arg \max_{1 \leq i \leq m} [\delta_{t-1}(i)P_{ij}] \text{ for } 2 \leq t \leq T, 1 \leq j \leq m \qquad (3.28)$$

3. Termination:

$$Pr^* = \max_{1 \leq i \leq m} [\delta_T(i)] \qquad (3.29)$$

$$x(T)^* = \arg \max_{1 \leq i \leq m} [\delta_T(i)] \qquad (3.30)$$

4. Path (state sequence) backtracking:

$$x(t)^* = \psi_{t+1}(q_{t+1}^*) \text{ for } t = T - 1, T - 2, \ldots, 1 \qquad (3.31)$$

3.1.1 Baum-Welch Algorithm

Finally, a method to adjust the model parameters (P, M, ϕ) to maximize the probability of the observation sequence given the model $Pr(Y|\lambda)$ is defined. Given any finite observation sequence as training data, there is no optimal way of estimating the model parameters. One solution is to choose $\lambda = (P, M, \phi)$ such that $Pr(Y|\lambda)$ is locally maximized using an iterative procedure such as the *Baum-Welch* method.

The variable $\xi_t(i, j)$ is defined as the probability of being in state S_i at time t, and state S_i, at time $t + 1$, given the model and the observation sequence:

$$\xi_t(i, j) = Pr(x(t) = S_i, x(t + 1) = S_j | Y, \lambda)$$

$$= \frac{\alpha_t(i)P_{ij}M_j(y(t + 1))\beta_{t+1}(j)}{Pr(Y|\lambda)}$$

$$= \frac{\alpha_t(i)P_{ij}M_j(y(t + 1))\beta_{t+1}(j)}{\sum_{i=1}^{m}\sum_{j=1}^{m}\alpha_t(i)P_{ij}M_j(y(t + 1))\beta_{t+1}(j)} \qquad (3.32)$$

where the numerator term is just $Pr(x(t) = S_i, x(t+1) = S_j, Y|\lambda)$ and the division by $Pr(Y|\lambda)$ gives the desired probability measure.

Since $\sum_{t=1}^{T-1} \gamma_t(i)$ represents the expected number of transitions from S_i, and $\sum_{t=1}^{T-1} \xi_t(i, j)$ the expected number of transitions from S_i to S_j, the two variables $\gamma_t(i)$ and $\xi_t(i, j)$ are related by summing over j:

$$\gamma_t(i) = \sum_{j=1}^{m} \xi_t(i, j). \qquad (3.33)$$

Using the concept of counting event occurrences, the estimated parameters are defined as follows:

$$\overline{\phi}_i = \gamma_1(i) \tag{3.34}$$

$$\overline{P}_{ij} = \frac{\sum_{t=1}^{T-1} \xi_t(i, j)}{\sum_{t=1}^{T-1} \gamma_t(i)} \tag{3.35}$$

$$\overline{M}_j(k) = \frac{\sum_{t=1, s.t. y(t)=\mu_k}^{T} \gamma_t(j)}{\sum_{t=1}^{T} \gamma_t(j)} \tag{3.36}$$

The model $\overline{\lambda}$ is more likely than the model λ in the sense that $Pr(Y|\overline{\lambda}) > Pr(Y|\lambda)$, i.e., we have found a new model $\overline{\lambda}$ from which the observation sequence is more likely to have been produced. If $\overline{\lambda}$ is iteratively used in the place of λ and repeat the reestimation calculation, the probability of Y being observed from the model can be improved, until some limiting point is reached. The final result of this reestimation procedure is called a *maximum likelihood* estimate of the HMM. It has to be highlighted that the forward-backward algorithm leads to local maxima only.

The Baum-Welch reestimation equations are essentially identical to the EM steps for this particular problem, and the stochastic constraints of the HMM parameters are automatically satisfied at each iteration:

$$\sum_{i=1}^{m} \overline{\phi}_i = 1 \tag{3.37}$$

$$\sum_{j=1}^{m} \overline{P}_{ij} = 1 \text{ for } 1 \le i \le m \tag{3.38}$$

$$\sum_{k=1}^{s} \overline{M}_j(k) = 1 \text{ for } 1 \le j \le m \tag{3.39}$$

3.1.2   Factorial HMM

In an HMM, information about the past is conveyed through a single discrete variable, e.g., the hidden state. A generalization of HMMs in which this state is factored into multiple state variables and is therefore represented in a distributed manner.

An HMM encodes information about the history of a time series in the value of a single multinomial variable, e.g., the hidden state, which can take on one of m discrete values. This multinomial assumption supports an efficient parameter estimation algorithm, the Baum-Welch algorithm, which considers each of the m settings of the hidden state at each time step.

An HMM with a *distributed* state representation let the model automatically decompose the state space into features that decouple the dynamics of the process that generated the data, therefore the task of modelling time series that are known a priori to be generated from an interaction of multiple, loosely-coupled processes.

The treatise followed in this chapter is inspired by Ghahramani and Jordan [77].

The generalization of the HMM state representation let the state be represented by a collection of state variables:

$$x(t) = \{x^{(1)}(t), \dots, x^{(i)}(t), \dots, x^{(N)}(t)\}, \tag{3.40}$$

where N is the number of underlying distributed variables, each of which can take on m_i values.

This model is defined as *Factorial Hidden Markov model* (FHMM), as the state space consists of the cross product of these state variables. The number of state combination is equal to $\prod_{i=1}^{N} m_i$.

A natural structure to consider is one in which each state variable evolves according to its own dynamics, and is *a priori* uncoupled from the other state variables:

$$Pr(x(t)|x(t-1)) = \prod_{i=1}^{N} Pr(x^{(i)}(t)|x^{(i)}(t-1)) \tag{3.41}$$

The observation at time step t can depend on all the state variables at that time step. For continuous observations, as a linear Gaussian, the observation $\overline{y}(t)$ is a random vector whose mean is a linear function of the state variables. Representing the state variables $\mathbf{x}^{(i)}(t)$ as $[m_i \times 1]$ vectors, where each of the m_i discrete values corresponds to a 1 in one position and 0 elsewhere. The probability density for a $[n \times 1]$ observation vector $\overline{y}(t)$:

$$Pr(\overline{y}(t)|x(t)) = |\mathbf{C}|^{-\frac{1}{2}}(2\pi)^{-\frac{n}{2}} \exp\left\{-\frac{1}{2}(\overline{y}(t) - \boldsymbol{\mu}(t))'\mathbf{C}^{-1}(\overline{y}(t) - \boldsymbol{\mu}(t))\right\}, \tag{3.42}$$

where

$$\boldsymbol{\mu}(t) = \sum_{i=1}^{N} \mathbf{W}^{(i)}\mathbf{x}^{(i)}(t). \tag{3.43}$$

Each $\mathbf{W}^{(i)}$ matrix is an $[n \times m_i]$ matrix whose columns are the contributions to the means for each of the settings of $\mathbf{x}^{(i)}(t)$, $\mathbf{C}$ is the $[n \times n]$ covariance matrix, ' denotes matrix transpose, and $|\cdot|$ is the matrix determinant operator.

The inference problem consists of computing the probabilities of the hidden variables given the observations. This problem can be solved efficiently via the forward-backward algorithm. In some cases, it is desirable to infer the single most probable hidden state sequence. This can be achieved via the Viterbi algorithm.

The learning problem consists of learning the parameters for a given structure. The parameters of a factorial HMM can be estimated via the Expectation Maximization (EM) algorithm, which in the case of classical HMMs is known as the Baum-Welch algorithm. This procedure iterates between a step that fixes the current parameters and computes posterior probabilities over the hidden states (the E step) and a step that uses these probabilities to maximize the expected log likelihood of the observations as a function of the parameters (the M step). The exact M step for factorial HMMs is simple and tractable, whilst the exact E step for factorial HMMs is computationally intractable. Rather than computing the exact posterior probabilities, one can approximate them using a Monte Carlo sampling procedure, avoid the sum over exponentially many state patterns at some cost in accuracy. Within many possible sampling schemes, the *Gibbs* sampling is the simplest. A second approach is the *Completely factorized variational* inference, which results to be both tractable and deterministic. A third approximation, the *Structured variational* inference, is both tractable and preserves much of the probabilistic structure of the original system.

3.2 Deep Neural Network (DNN)

A *biological Neural Networks* is a big set of specialized cells (*neurons*) connected among them, which memorize and process information, thus controlling the body activities they belong to.

The *neuron* model is composed of:

- Dendrite, as the input terminal.
- Cell body (Nucleus), as the processing core.
- Axon, as the output way-out.
- Synapses, as the output terminal (with weight).

The *neuron* properties can be described in:

- Local simplicity, since the neuron receives stimuli (excitation or inhibition) from dendrites and produces an impulse to the axon which is proportional to the weighted sum of the inputs.
- Global complexity, since the human brain possesses $\mathcal{O}(10^{10})$ neurons, with more than 10K connections each.
- Learning, since the strength of connections (synaptic weights) can change when the network is exposed to external stimuli, even though the network topology is relatively fixed.
- Distributed control, since each neuron reacts only to its own stimuli.

- Tolerance to failures, since performance slowly decreases with the increase of failures.

The biological Neural Networks are able to solve very complex tasks in few time instants (like memorization, recognition, association and so on.)

The *Artificial Neural Networks* (ANNs) are defined as *Massively parallel distributed processors made up of simple processing units having a natural propensity for storing experiential knowledge and making it available for use* [78].

An ANN resembles the brain in two aspects:

1. Knowledge is acquired by the network from its environment through a learning process.
2. Synaptic weights are used to store the acquired knowledge.

A *neuron* is an information-processing unit that is fundamental to the operation of a neural network. The model of a neuron is composed of three basic elements of the neural model:

- A *set of synapses*, or connecting links, each of which is characterized by a weight or strength of its own, w_{kj}.
- An *adder* for summing the input signals, weighted by the respective synaptic strengths of the neuron; the operations described here constitute a linear combiner.
- An *activation function* for limiting the amplitude of the output of a neuron. Typically, the normalized amplitude range of the output of a neuron is written as the closed unit interval [0,1], or, alternatively, [-1,1].

The neural model also includes an externally applied *bias*, denoted by b_k.

Therefore, the mathematical description of neuron activity can be defined as:

$$u_k = \sum_{j=1}^{m} w_{kj} x_j \tag{3.44}$$

$$y_k = \varphi \left(u_k + b_k \right) \tag{3.45}$$

where:

- $x_1, x_2, \cdots, x_m$ are the input signals.
- $w_{k1}, w_{k2}, \cdots, w_{km}$ are the respective synaptic weights of neuron k.
- u_k is the linear combiner output due to the input signals.
- b_k is the bias.
- $\varphi(\cdot)$ is the activation function.
- y_k is the output signal of the neuron.

The types of *non-linear activation functions* $\varphi(v)$ are:

- The *threshold function*, commonly referred to as a Heaviside function.

$$\varphi(v) = 1 \quad if \quad v \geq 0, \tag{3.46}$$

$$\varphi(v) = 0 \quad if \quad v < 0. \tag{3.47}$$

- The *sigmoid function*, which is defined as a strictly increasing function that exhibits a graceful balance between linear and non-linear behaviour. An example of the sigmoid function is the *logistic function*.

$$\varphi(v) = \frac{1}{1 + exp(-av)} \tag{3.48}$$

- The *hyperbolic tangent* ($tanh$), which is simply a scaled and shifted version of the sigmoid function.

$$\varphi(v) = \frac{1 - e^{-2v}}{1 + e^{-2v}} \tag{3.49}$$

- The *Rectifier Linear Unit (ReLU)*.

$$\varphi(v) = \max(0, v) \tag{3.50}$$

- The *softmax*, which is used on the last layer of a classifier setup: the outputs of the softmax layer represent the probabilities that a sample belongs to the different classes. Indeed, the sum of all the output is equal to 1.

$$\varphi(v_k) = \frac{e^{v_k}}{\sum_{j=1}^{K} e^{v_j}} \quad \text{for } k = 1, \ldots, K \tag{3.51}$$

The manner in which the neurons of a neural network are structured is intimately linked with the learning algorithm used to train the network. There, the *network architectures* (structures) are defined. In general, two different classes of network architectures are identified: the *Multilayer Feed-forward Networks* (FFNN) and the *Convolutional Neural Networks* (CNN).

The Multilayer Feed-forward Networks is characterized by the presence of one or more hidden layers, whose computation nodes are correspondingly called *hidden neurons* (or hidden units). The term *hidden* refers to the fact that this part of the neural network is not seen directly from either the input or output of the network. The function of hidden neurons is to intervene between the external input and the network output in some useful manner. By adding one or more hidden layers, the network is enabled to extract higher-order statistics from its input. The MLP is a well-known kind of artificial neural network introduced in 1986 [79]. Each node applies an activation function over the weighted sum of its inputs. The units are arranged in layers, with feed-forward connections from one layer to the next. The stochastic gradient descent with error back-propagation algorithm is used for the supervised learning of the network. In the forward pass, input examples are fed to the input layer, and the resulting output is propagated via the hidden layers towards the output layer. At the backward pass, the error signal originating at the output neurons is sent back through the layers and the network parameters (i.e., weights and biases) are tuned. A single neuron can be formally described as:

$$g(\mathbf{x}) = \varphi \left(\sum_{j=1}^{m} w_j x_j + b \right), \tag{3.52}$$

where $\mathbf{x} \in \mathbb{R}^{m \times 1}$, the bias b is an externally applied term and $\varphi(\cdot)$ is the non-linear activation function. Thus, the mathematical description of a one-hidden-layer MLP is a function $f : \mathbb{R}^m \rightarrow \mathbb{R}^{m'}$, where m' is the size of the output vector $\mathbf{y}$, so:

$$\mathbf{y} = f(\mathbf{x}) = \varphi \left(\mathbf{b}_2 + \mathbf{W}_2 \left(\varphi \left(\mathbf{b}_1 + \mathbf{W}_1 \cdot \mathbf{x} \right) \right) \right), \tag{3.53}$$

where $\mathbf{W}_i$ and $\mathbf{b}_i$ are the respective synaptic weight matrix and the bias vector of the i-th layer. The behaviour of this architecture is parametrized by the connection weights, which are adapted during the supervised network training.

The Convolutional neural networks are feed-forward neural networks similar to multilayer perceptron, with some special layers. Convolution kernels process the input data matrix by dividing it in *local receptive fields*, a region of the same size of the kernel, and sliding the local receptive field across the entire input. Each hidden neuron is thus connected to a local receptive field, and all the neurons form a matrix called *feature map*. The weights in each *feature map* are *shared*: all hidden neurons are aimed to detect exactly the same pattern just at different locations in the input image. The main advantage of this network is the robust pattern recognition system characterized by a strong immunity to pattern shifts. Pooling layer just reduces the dimension of the matrix by a rule: a submatrix of the input is selected, and the output is the maximum value of this submatrix. The pooling process introduces tolerance against shifts of the input patterns. Together with convolution layer it allows the CNN to detect if a particular event occurs, regardless of its deformation or its position. CNN is a feed-forward neural network [80] usually composed of three types of layers: convolutional layers, pooling layers and layers of neurons. The convolutional layer performs the mathematical operation of convolution between a multi-dimensional input and a fixed-size kernel. Successively, a non-linearity is applied element-wise. The kernels are generally small compared to the input, allowing CNNs to process large inputs with few trainable parameters. Successively, a pooling layer is usually applied, in order to reduce the feature map dimensions. One of the most used is the *max-pooling* whose aim is to introduce robustness against translations of the input patterns. Finally, at the top of the network, a layer of neurons is applied. This layer does not differ from MLP, being composed by a set of activation and being fully connected with the previous layer. For clarity, the units contained in this layer will be referred to as *Hidden Nodes* (HN). Denoting with $\mathbf{W}_k \in \mathbb{R}^{k_1 \times k_m}$ the k-th kernel and with $\mathbf{b}_k \in \mathbb{R}^{m_1 \times m_2}$ the bias vector of a generic convolutional layer, the k-th feature map $\mathbf{h}_k \in \mathbb{R}^{m_1 \times m_2}$ is given by:

$$\mathbf{h}_k = \varphi \left(\sum_{d=1}^{m_3} \mathbf{W}_k * \mathbf{u}_m + \mathbf{b}_k \right), \tag{3.54}$$

where $*$ represents the convolution operation, and $\mathbf{u}_m \in \mathbb{R}^{m_1 \times m_2}$ is a matrix of the three-dimensional input tensor $\mathbf{u} \in \mathbb{R}^{m_1 \times m_2 \times m_3}$. The dimension of the k-th feature map $\mathbf{h}_k$ depends on the zero padding of the input tensor: here, padding is performed in order to preserve the dimension of the input, i.e., $\mathbf{h}_k \in \mathbb{R}^{m_1 \times m_2}$. Commonly, (3.54) is followed by a pooling layer in order to be more robust against patterns shifts in the processed data, e.g. a max-pooling operator that calculates the maximum over a $p_1 \times p_2$ matrix is employed.

The *Deep Learning* is a class of machine learning techniques that exploits many layers of non-linear information processing for supervised or unsupervised feature extraction and transformation, and for pattern analysis and classification. Artificial Neural Networks are often referred to as deep when they have more than one or two hidden layers.

3.2.1 Stochastic Gradient Descent (SGD)

Most deep learning training algorithms involve optimization of some sort. The most widely used is the gradient based optimization, which belongs to the first order type. The treatise followed in this chapter is inspired by Goodfellow et al. [81].

Optimization is the task of minimizing some function $f(x)$ by altering x: $f(x)$ is called *objective function*, but in the case when it has to be minimized, it is also call the *cost function*, *loss function* or *error function*. The aim of the optimization is reached doing small change ϵ in the input x, to obtain the corresponding change in the output $f(x)$:

$$f(x + \epsilon) \approx f(x) + \epsilon f'(x). \tag{3.55}$$

This formulation is based on the calculation of the derivative $f'(x)$. The *gradient descent* is the technique based on the reduction of $f(x)$ by moving x in small steps with the opposite sign of the derivative. The aim is to find the minimum of the cost function: when $f'(x) = 0$, the derivative provides no information about which direction to move, therefore this point is defined as stationary points. A local minimum is a point where $f(x)$ is lower than at all neighbouring and it is no longer possible to decrease $f(x)$ by making infinitesimal steps. The absolute lowest value of $f(x)$ is a *global minimum*.

For the concept of minimization to make sense, there must still be only one (scalar) output. For functions that have multiple inputs $f : \mathbb{R}^n \rightarrow \mathbb{R}$, the concept of *partial derivatives* is introduced. The gradient $\nabla_{\mathbf{x}} f(\mathbf{x})$ is the vector containing all the partial derivatives.

The method of steepest descent or gradient descent states that f decreases by moving in the direction of negative gradient.

$$\mathbf{x}' = \mathbf{x} - \epsilon \, \nabla_{\mathbf{x}} f(\mathbf{x}), \tag{3.56}$$

where ϵ is the *learning rate*, a positive scalar determining the size of the step.

Large training sets are necessary for good generalization, but large training sets are also more computationally expensive. The cost function decomposes as a sum over training example of per-example loss function: i.e., the negative conditional log-likelihood of the training data is defined as:

$$J(\boldsymbol{\theta}) = \mathbb{E}(L(\mathbf{x}, y, \boldsymbol{\theta})) = \frac{1}{m} \sum_{i=1}^{m} L(\mathbf{x}^{(i)}, y^{(i)}, \boldsymbol{\theta}), \tag{3.57}$$

where L is the per-example loss $L(\mathbf{x}, y, \boldsymbol{\theta}) = -\log p(y|\mathbf{x}; \boldsymbol{\theta})$. The gradient descent requires computing:

$$\nabla_{\theta} J(\boldsymbol{\theta}) = \frac{1}{m} \sum_{i=1}^{m} \nabla_{\theta} L(\mathbf{x}^{(i)}, y^{(i)}, \boldsymbol{\theta}). \tag{3.58}$$

The computational cost of this operation is proportional to the number of example m, therefore as the training set size grows the time to take a single gradient step becomes prohibitively long.

Stochastic gradient descent (SGD) is an extension of the gradient descent algorithm: the insight is that the gradient is an expectation estimated using a small set of samples. On each step of the algorithm, a sample of example $\mathbb{B} = \{\mathbf{x}^{(1)}, \ldots, \mathbf{x}^{(m')}\}$, called *minibatch*, is drawn uniformly from the training set. The minibatch size m' is typically chosen to be a relatively small number of examples. The estimate of the gradient is: $\mathbf{g} = \frac{1}{m'} \nabla_{\theta} \sum_{i=1}^{m'} L(\mathbf{x}^{(i)}, y^{(i)}, \boldsymbol{\theta})$ using examples from the minibatch $\mathbb{B}$. The SGD algorithm then follows the estimated gradient downhill:

$$\theta \leftarrow \theta - \epsilon \, \mathbf{g} \tag{3.59}$$

where ϵ is the learning rate.

3.2.2 Autoencoder

An autoencoder is a kind of neural network typically consisting of only one hidden layer, trained to set the target values to be equal to the inputs.

Given an input set of examples $\mathcal{X}$, autoencoder training consists in finding parameters θ that minimize the reconstruction error.

Defining h the number of hidden units, and m the number of input units, output units, features size:

- $h = m \rightarrow$ Basic Autoencoder (AE);
- $h < m \rightarrow$ Compression Autoencoder (CAE);
- $h > m$ and Gaussian Noise $\rightarrow$ Denoising Autoencoder (DAE);

3.2.2.1 Basic Autoencoder

A basic AE—a kind of neural network typically consisting of only one hidden layer—sets the target values to be equal to the input. It is used to find common data representation from the input [82, 83]. Formally, in response to an input example $\mathbf{x} \in \mathbb{R}^m$, the hidden representation $h(\mathbf{x}) \in \mathbb{R}^h$ is defined:

$$h(\mathbf{x}) = f(\mathbf{W_1}\mathbf{x} + \mathbf{b_1}), \tag{3.60}$$

where $f(z)$ is the non-linear activation function applied component-wisely, $\mathbf{W_1} \in \mathbb{R}^{h \times m}$ is a weight matrix and $\mathbf{b_1} \in \mathbb{R}^h$ is a bias vector.

The network output maps the hidden representation $h(\mathbf{x})$ back to a reconstruction $\mathbf{y} \in \mathbb{R}^m$:

$$\mathbf{y} = f(\mathbf{W_2}h(\mathbf{x}) + \mathbf{b_2}), \tag{3.61}$$

where $\mathbf{W_2} \in \mathbb{R}^{m \times h}$ is a weight matrix and $\mathbf{b_2} \in \mathbb{R}^m$ is a bias vector.

Given an input set of examples $\mathcal{X}$, AE training consists in finding parameters $\theta = \{\mathbf{W_1}, \mathbf{W_2}, \mathbf{b_1}, \mathbf{b_2}\}$ that minimize the reconstruction error, which corresponds to minimizing the following objective function:

$$\mathcal{J}(\theta) = \sum_{\mathbf{x} \in \mathcal{X}} \|\mathbf{x} - \mathbf{y}\|^2. \tag{3.62}$$

The minimization is usually realized by stochastic gradient descent as in the training of neural networks.

3.2.2.2 Compression Autoencoder

In the case of having the number of hidden units h smaller than the number of input units m, the network is forced to learn a compressed representation of the input. For example, if some of the input features are correlated, then this compression autoencoder (CAE) is able to learn those correlations and reconstruct the input data from a compressed representation.

3.2.2.3 Denoising Autoencoder

The denoising AE (DAE) [84] forces the hidden layer to retrieve more robust features and prevent it from simply learning the identity. In such a configuration the AE is trained to reconstruct the original input from a corrupted version of it. Formally, the initial input $\mathbf{x}$ is corrupted by means of additive isotropic Gaussian noise in order to obtain: $\mathbf{x}'|\mathbf{x} \sim N(\mathbf{x}, \sigma^2 \mathbf{I})$. The corrupted input $\mathbf{x}'$ is then mapped, as with the AE, to a hidden representation, defined as:

$$h(\mathbf{x}') = f(\mathbf{W}_1'\mathbf{x}' + \mathbf{b}_1'), \tag{3.63}$$

from which the original signal is reconstructed as follows:

$$\mathbf{y} = f(\mathbf{W}_2' h(\mathbf{x}') + \mathbf{b}_2'). \tag{3.64}$$

The parameters $\theta' = \{\mathbf{W}_1', \mathbf{W}_2', \mathbf{b}_1', \mathbf{b}_2'\}$ are trained to minimize the average reconstruction error over the training set, to have $\mathbf{y}$ reach as close as possible to the uncorrupted input $\mathbf{x}$, which corresponds to minimizing the objective function.

Chapter 4
HMM Based Approach

Abstract Approaches based on hidden Markov models (HMMs) have been devoted particular attention in the last years. AFAMAP (Additive Factorial Approximate Maximum a Posteriori) has been introduced in Kolter and Jaakkola to reduce the computational burden of FHMM. The algorithm bases its operation on additive and difference FHMM, and it constrains the posterior probability to require only one HMM change state at any given time.

Keywords Hidden Markov Model · Working state · Power consumption · Active power · Factorial Hidden Markov Model · Rest-of-the-world model · Constrained optimization · Reactive power · Finite State Machine · Footprint

Each appliance is modelled as an n-variate HMM, i.e., an HMM whose emitted symbols are represented by n values. More in details, each HMM is represented by the following parameters [76]:

- the number of states $m \in \mathbb{Z}_+$;
- the hidden states $x \in \{S_1, S_2, \ldots, S_m\}$;
- the symbols emitted $\boldsymbol{\mu}_j \in \mathbb{R}^n$, where $j = 1, \ldots, s$;
- the symbol emission probability matrix $\boldsymbol{M}^{s \times m}$;
- the state transition probability matrix $\boldsymbol{P} \in [0, 1]^{m \times m}$;
- the starting state probability vector $\boldsymbol{\phi} \in [0, 1]^m$.

In this book, it is assumed that each state of the HMM corresponds to a working state of the appliance, i.e., $x \in \{ON_1, ON_2, \ldots, OFF\}$, so that the number of states m is equal to the number of symbols s and $\boldsymbol{M} \equiv \boldsymbol{I}^{m \times m}$ (*degenerate HMM*). Furthermore, the values composing the emitted symbols represent the power consumption of the appliance: since the components are defined in an orthogonal space, the power consumptions which best fit with this constraint are the active and reactive power. Therefore, $n = 2$ and for the sake of clarity in the remainder of this section it will be omitted since the individual active and reactive power components will be made explicit. For example, each symbol is defined as $\boldsymbol{\mu}_j = [\mu_{a,j} \, \mu_{r,j}]^T$, where the subscripts a and r distinguish the active and reactive components. In

R. Bonfigli, S. Squartini, *Machine Learning Approaches to Non-Intrusive Load Monitoring*, SpringerBriefs in Energy, https://doi.org/10.1007/978-3-030-30782-0_4

the remainder of this section, there will be additionally treated the analysis in the unidimensional space, with $n = 1$, exploiting only one of the two components. For each appliance, the quantities to be estimated are the number of states m, the values of μ_j for each state, the state transition probability matrix P and the starting state probability vector ϕ. Estimation of m and of μ_j will be addressed in Sect. 4.1.1.

Regarding the state transition probability matrix P, each entry P_{ij} represents the probability of transitioning from state i to state j. Thus, P_{ij} can be estimated with a Maximum Likelihood criterion by calculating the number of times state i transitions to state j and normalizing by the total number of transitions from state i. Formally:

$$P_{ij} = \frac{C_{ij}}{\sum_{j'=1}^{m} C_{ij'}}, \tag{4.1}$$

where C_{ij} is the number of transitions from state i to state j. Typically, the greatest values in the matrix are located in the diagonal, meaning that the probability of remaining in the same state is higher compared to the probability of transitioning to another state. Table 4.1 shows a typical transition probability matrix related to an appliance with four working states. The highest values in the matrix are the ones located on the diagonal, which represent the probability of remaining in the same state, with respect to the transition to another one: indeed, for the states where the permanence time is low, this value is lower than the one of the state where the permanence time is higher. The highest value is the one related to the OFF state, because the activation of the appliance occurs after a long time in which it is turned off.

In addition, the OFF state corresponds to the initial state, since the footprint starts just before the turning on instant, thus $\phi = [0\,0\cdots0\,1]^T$.

An example of a four-state appliance model is shown in Fig. 4.1, where the arc between two states is the probability of transition P_{ij}, while the arc starting and closing on the same state represents the probability P_{ii} of permanence in each state.

The probability value which tends to zero denotes that the transition is unlikely. In practice, it is recommended to avoid zero probability value, because it is evaluated in log scale in the AFAMAP algorithm, and it tends to infinity. It is recommended to fix the value to a little quantity, e.g., $\simeq 10^{-5}$.

Table 4.1 An example of the HMM transition probability matrix

	Destination state			
Start state	ON$_1$	ON$_2$	ON$_3$	OFF
ON$_1$	0.832	0.085	0.081	0.002
ON$_2$	0.080	0.690	0.202	0.028
ON$_3$	0.012	0.028	0.916	0.045
OFF	3.1e$-$05	2.7e$-$05	0.002	0.998

Fig. 4.1 An example of a
four-state HMM

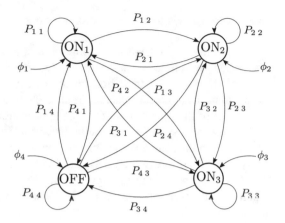

4.1   Additive Factorial Approximate Maximum A Posteriori (AFAMAP)

FHMMs have been introduced in [77] as an extension of HMMs to model time series that depend on multiple hidden processes. Starting from the work of Kim and colleagues [17], FHMMs have been largely employed for NILM and several approaches have been proposed in the literature [18, 22, 24, 27, 85, 86]. Among them, AFAMAP [21] represents an effective algorithm able to achieve high performance with a reasonable computational cost.

AFAMAP has been proposed in [21] as an efficient disaggregation algorithm based on FHMMs. In this algorithm, an additional model which relies on the same HMMs composing the Additive FHMM (AFHMM) is introduced. It is based on a differential version of the aggregated signal, resulting in a Differential FHMM (DFHMM). The inference on the set of states of multiple HMMs can be computed through the Maximum A Posteriori (MAP) algorithm and a relaxation towards real values is taken into account, leading to a convex Quadratic Programming (QP) optimization problem. The disaggregation process is performed by analyzing the aggregated power divided in non-overlapping frames.

The reference work [21] describes an unsupervised approach to data disaggregation: in fact, an unsupervised procedure aimed to the extraction of the device load signature is paired with the disaggregation algorithm, referred to as AFAMAP (Additive Factorial Approximate Maximum a Posteriori). In this work, the aim is to investigate and to improve the disaggregation algorithm. Differently to the reference work, however, a supervised approach is used to create the HMMs, based on the circuit level power consumption signature. The signal can be obtained, clearly, from the aggregated data under the condition that the appliances run one at a time [15].

The theoretical approach towards disaggregation is based on the work of Kolter and Jaakkola [21]. In this work the system is modelled relying on Additive Factorial Hidden Markov Model (AFHMM), for which the value of each aggregated power sample corresponds to a combination of working states of the appliances into the system.

Also, in this approach, the assumption that at most one HMM may change its state at any given time is made, which holds true if the sampling time is reasonably short. In this case, the transition on the aggregate power, when moving from a sample to the next, corresponds to the state change of a particular HMM.

Because of that, the differential signal, built from the aggregated power, can be modelled as the result of a Differential Factorial Hidden Markov Model (DFHMM), which relies on the same HMM models comprising the AFHMM.

By combining the two models, the inference on the set of states of multiple HMMs can be computed through the Maximum A Posteriori (MAP) technique, which takes the form of a Mixed Integer Quadratic Programming (MIQP) optimization problem.

One of the shortcomings of this approach is the non-convex nature of the problem, because of the integer nature of the variables: in this case, a relaxation towards real values is taken into account, leading to a convex Quadratic Programming (QP) optimization problem. Thus, the Additive Factorial Approximate MAP (AFAMAP) approach is obtained.

In a real case scenario, the modelled output may not match with the observed aggregated signal, due to electrical noises, very small loads, or leakages. In that case, the issue is addressed by defining a robust mixture component in both AFHMM and DFHMM, named $z(t)$ and $\Delta z(t)$, respectively.

When a *denoised* scenario [87] is considered, i.e., all the contributions to the aggregated energy demand are known, the robust mixture component is missing. When a *noised* scenario is considered, the robust mixture component is not used, and all the contributions are modelled as an additional appliance represented by the RoW model, which will be introduced in Sect. 4.1.2. This approach provides further advantages, since appliances with lower power consumption values risk to be modelled with working states associated to similar consumption values. This can lead the algorithm to an erroneous assignment of the disaggregation output between similar models. Furthermore, the authors in [21] demonstrated that the disaggregation performance degrades as the number of appliances increases. Thus, representing several appliances with a single model eases the disaggregation task.

In the reference work [21], the parameter n defines the problem dimensionality: in its presentation, it is assumed $n = 1$, because the algorithm uses only the active power data to characterize the observed aggregated signal.

Specifically, the parameters of the problem follows:

- $N \in \mathbb{Z}_+$ is the number of HMMs in the system;
- $\overline{y}(\tau) \in \mathbb{R}$ is the observed aggregated output (in denoised environments $\overline{y}(\tau) = \sum_{i=1}^{N} y^{(i)}(\tau)$, where $y^{(i)}(\tau)$ corresponds to the true appliance output);
- $\sigma_{1/2}^2 \in \mathbb{R}$ is the observation variance.

The differential signal is referred to as $\Delta \overline{y}_b(\tau) = \overline{y}(\tau) - \overline{y}(\tau - 1)$.

For the i-th HMM the parameters are:

- $m_i \in \mathbb{Z}_+$ is the number of states;
- $x^{(i)}(\tau) \in \{S_1, \ldots, S_{m_i}\}$ is the HMM state at time instant τ ($x^{(i)}(\tau) \equiv S_{m_i}$ corresponds to the OFF state);
- $\mu_j^{(i)} \in \mathbb{R}$ is the j-th state mean value;
- $\phi_b^{(i)} \in [0, 1]^{m_i}$ is the initial states distribution;
- $P_b^{(i)} \in [0, 1]^{m_i \times m_i}$ is the transition matrix.

The aggregated signal $\overline{y}(\tau)$ is analysed using a windowing technique, where $\tau \in w_f = [(f-1)T+1, \ldots, fT]$ for $f = 1, 2, \ldots, F$. The window w_f is the timebase for the algorithm and, for convenience, a new temporal variable is introduced by defining the relation $t = \tau - (f-1)T$, for $t = 1, 2, \ldots, T$, with $T \in \mathbb{Z}_+$. After the analysis of all the F windows, the disaggregated signals $\hat{y}^{(i)}(t)$ are recomposed using the inverse relation $\tau = t + (f-1)T$.

In the optimization problem, the variables are defined as:

$$\mathcal{Q} = \left\{ Q(x^{(i)}(t)) \in \mathbb{R}^{m_i}, \; Q(x^{(i)}(t-1), x^{(i)}(t)) \in \mathbb{R}^{m_i \times m_i} \right\},$$

for which the $Q(x^{(i)}(t))_j$ variable is the indicator of the state assumed at time instant t, while the $Q(x^{(i)}(t-1), x^{(i)}(t))_{j,k}$ variable is the indicator of the state transition from previous to actual time instant, for the i-th HMM.

The AFAMAP algorithm is shown in Fig. 4.2.

Input: $\overline{y}(1 : T)$ aggregated signal; $\left\{ \mu^{(1:N)}, P_b^{(1:N)}, \phi_b^{(1:N)} \right\}$ parameters for N HMMs; $\sigma_1^2, \sigma_2^2, \lambda$ covariance and regularization parameters.

Minimize over $\{\mathcal{Q} \in \mathcal{L} \cap \mathcal{O}\}$

$$\frac{1}{2\sigma_1^2} \sum_{t=1}^{T} E'(t) + \frac{1}{2\sigma_2^2} \sum_{t=2}^{T} E''(t) + \frac{1}{2} \sum_{t=2}^{T} E'''(t) +$$

$$+ \sum_{t=2}^{T} \sum_{i=1}^{N} \sum_{\substack{j=1 \\ k=1}}^{m_i} \left\{ Q(x^{(i)}(t-1), x^{(i)}(t))_{j,k} \left(-\log P_{b_{k,j}}^{(i)} \right) \right\} + \quad (4.2)$$

$$+ \sum_{i=1}^{N} \sum_{j=1}^{m_i} \left\{ Q(x^{(i)}(1))_j \left(-\log \phi_{b_j}^{(i)} \right) \right\}$$

Output : $\hat{y}^{(1:N)}(1 : T)$, predicted individual HMM output

$$\hat{y}^{(i)}(t) = \sum_{j=1}^{m_i} \mu_j^{(i)} Q(x^{(i)}(t))_j \quad (4.3)$$

Fig. 4.2 The AFAMAP algorithm

In (4.2) the error terms are defined as:

$$E'(t) = \left(\bar{y}(t) - \sum_{i=1}^{N}\sum_{j=1}^{m_i}\left\{\mu_j^{(i)}Q(x^{(i)}(t))_j\right\}\right)^2, \tag{4.4}$$

$$E''(t) = \sum_{\substack{i=1 \\ }}^{N}\sum_{\substack{j=1 \\ k=1 \\ k\neq j}}^{m_i}\left\{\left(\Delta\bar{y}_b(t) - \Delta\mu_{k,j}^{(i)}\right)^2 Q(x^{(i)}(t-1), x^{(i)}(t))_{j,k}\right\}, \tag{4.5}$$

$$E'''(t) = D\left(\frac{\Delta\bar{y}_b(t)}{\sigma_2}, \lambda\right)\left(1 - \sum_{\substack{i=1 \\ }}^{N}\sum_{\substack{j=1 \\ k=1 \\ k\neq j}}^{m_i} Q(x^{(i)}(t-1), x^{(i)}(t))_{j,k}\right). \tag{4.6}$$

The QP optimization problem is defined in the form:
Minimize

$$\frac{1}{2}v^T H v + f^T v, \tag{4.7}$$

subject to the constraint:

$$A_{eq}v = b_{eq}, \tag{4.8}$$

$$lb \leq v \leq ub. \tag{4.9}$$

The variables of the problem are represented by the vector v, which is composed of several subsets, based on the time instant t and the appliance index (i):

$$v = \begin{bmatrix} \Theta(1) \\ \vdots \\ \Theta(T) \end{bmatrix}, \quad \Theta(t) = \begin{bmatrix} \Psi^{(1)}(t) \\ \vdots \\ \Psi^{(N)}(t) \end{bmatrix}, \quad \Psi^{(i)}(t) = \begin{bmatrix} \xi^{(i)}(t) \\ \beta^{(i)}(t) \end{bmatrix},$$

$$\xi^{(i)}(t) = \begin{bmatrix} Q(x^{(i)}(t))_1 \\ \vdots \\ Q(x^{(i)}(t))_{m_i} \end{bmatrix}, \quad \beta^{(i)}(t) = \begin{bmatrix} Q(x^{(i)}(t-1), x^{(i)}(t))_{1,1} \\ \vdots \\ Q(x^{(i)}(t-1), x^{(i)}(t))_{1,m_i} \\ \vdots \\ Q(x^{(i)}(t-1), x^{(i)}(t))_{m_i,1} \\ \vdots \\ Q(x^{(i)}(t-1), x^{(i)}(t))_{m_i,m_i} \end{bmatrix},$$

where the variables for the state are represented in $\xi^{(i)}(t)$, and the variables for the backward transition in $\beta^{(i)}(t)$.

Fig. 4.3 Additive FHMM
model

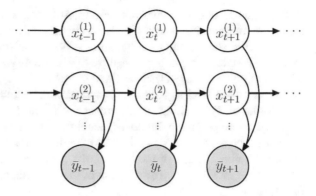

Fig. 4.4 Differential FHMM
model

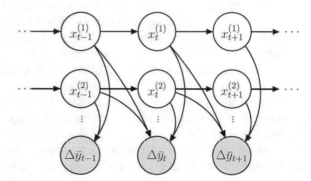

The parameters of the problem fill up the elements of H and f, according to the
structure of the v vector, whereas A_{eq} and b_{eq} are used to represent the consistent
constraints between the state and the transition variables. The vectors lb and ub
define the lower and upper boundaries of the solution: because of the nature of the
variables [21], the lower boundary is equal to 0, whereas the upper boundary to 1,
for all the elements in v (Fig. 4.3).

In A_{eq} the constraint about $Q(x^{(i)}(t-1), x^{(i)}(t))$ with $t = 1$ has to be removed
since there is no information about $Q(x^{(i)}(t))$ at the previous time instant, thus
falling back to the constraint $0 \cdot Q(x^{(i)}(t-1), x^{(i)}(t)) = 0$ (Fig. 4.4).

4.1.1   Appliance Modelling

The working states power level estimation consists in obtaining representative
power level distributions related to each appliance state, i.e., the values of the
emitted symbols μ_j. In a realistic scenario, this is obtained by using a set of
examples of an appliance typical consumption cycle. This information can be
extracted by observing the aggregate power signal, under the assumption that only
one appliance at time is operating [15].

In particular, this stage involves the extraction of a *footprint* of the appliance, i.e., the active and reactive power signals comprised between the power on (transition from the OFF state to an ON state) and the power off (transition from an ON state to the OFF state). This is performed by firstly identifying these instants by means of an *Appliance Activity Detector* (AAD). Basically, it consists in detecting when the active power level signal exceeds a certain threshold or not (typical values are in the order of 20 W). Isolated occurrences of power levels below the threshold are managed by employing a *hangover* technique: it is a counter, which decreases its value for each sample the signal is below the threshold. If the signal returns over the threshold before the end of the counter, the footprint is considered continued. The typical value is 5–10 min. The diagram of the footprint extraction stage is shown in Fig. 4.5a.

The power value and the temporal information of the OFF state cannot be obtained by analysing the signal extracted with the AAD. The value is reasonably assumed 0 W and 0 VAR for the active and reactive power signals, respectively. The temporal information, i.e., the typical interval intercurring between the OFF state and an ON state, has to be specified a-priori for each appliance based on the typical usages (e.g., once in an hour, three times in a day, etc.).

Different uses of the appliance in its life cycle from the user lead to the need of model representation of every combination of usage, under the assumption that the working state of the appliance are predetermined and not varying from different usage: reasonably, the working cycle of a washing machine is always the same (e.g., pre-washing, water heating, washing, rinsing and spinning), indifferently from the

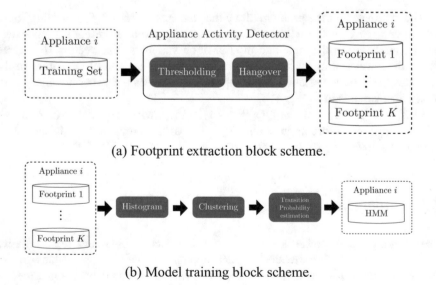

(a) Footprint extraction block scheme.

(b) Model training block scheme.

Fig. 4.5 Diagram of the footprint extraction procedure (**a**) and of the training phase of the appliance models (**b**)

order of execution, thus the number of working state is predetermined for every appliance.

Complex appliances (e.g., washing machines, dishwashers) are characterized by several working cycles and the extraction of a single footprint might not be completely representative of its operation. This motivates the need to acquire several footprints for each appliance. Furthermore, even though only one footprint is sufficient to explore all the working states of an appliance, multiple footprints allow to employ more data for the power level extraction phase, particularly useful for those power levels characterized by a short duration.

The estimation of the power level associated to a state of the HMM relies on the appliance consumption data, which is not composed of discrete values of consumption, but it presents a continuous variability in the values. In order to find the averaged values of the signal, within the period of permanence in the same working state, a clustering procedure is adopted, and the k-means [88] has been selected as the algorithm.

Since the OFF state information is not present in the data, the number of clusters is set to $(m - 1)$. After identifying the clusters, the power levels associated to each HMM state are represented by their centroids.

The clustering operation is not directly performed on the footprints extracted with the AAD. Indeed, after extracting the footprint, a bivariate histogram composed of 100 bins per kW and per kVAR is used to analyse the probability distribution of the active and reactive power signals. The number of bins is empirically chosen after analysing some footprints of the training set in order to obtain a sufficiently detailed histogram able to provide a good trade-off between variance and bias of the density estimate. Additionally, power levels with a low number of occurrences are excluded from the successive processing. More in details, bins having a number of occurrences below the threshold are considered of lower relevance, thus the related observations are discarded. This technique allows to obtain the number of working states m, which is determined by observing the number of clusters obtained in the final bivariate histograms. An example is shown in Fig. 4.6, where the histogram before and after the thresholding operation is shown. It refers to the dishwasher consumption in the AMPds dataset. Additionally, it reduces a limitation of the clustering algorithm: k-means does not employ the information on the samples distribution in the cluster, since it selects the centroid which satisfies the rule of convergence over all data. Discarding bins with low occurrences forces k-means to select the centroids with higher probability and to discard local clusters with lower probability, that could result in erroneous centroids. Furthermore, it allows to discriminate close clusters which can be confused as a single one: indeed, transients between near clusters produce samples comprised between the cluster with higher occurrences, which merge the two clusters in a single one.

Figure 4.7 represents the relationship between the cluster obtained on the consumption values in the footprints, and the footprint itself. It is related to the washing machine of the household 1, in the ECO dataset. In this case, the univariate modelling case, representing the active power consumption, is considered. The histogram, depicted in Fig. 4.7a, represents the probability density function of the

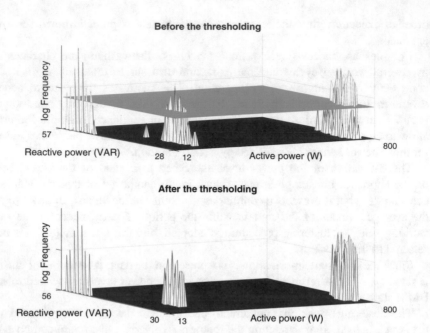

Fig. 4.6 An example of a two-dimensional histogram of the active and reactive power signals related to the dishwasher in the dataset AMPds

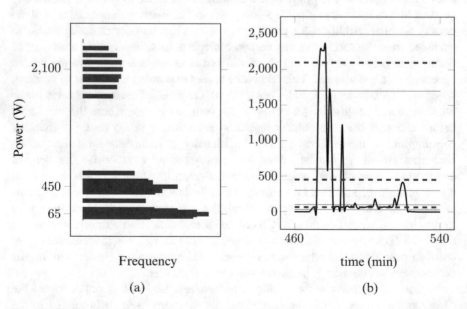

Fig. 4.7 Washing machine in ECO, household 1. (**a**) Histogram of the power consumption values. (**b**) Footprint and clusters associated to the working states

samples belonging to each state, which is correlated to the mean consumption value
of the same state and its variability, as shown in Fig. 4.7b.

The diagram of the clustering and of the model training stage is shown in
Fig. 4.5b.

In general, clusters present different characteristics depending on the magnitude
of their centroid. Typically, the ones characterized by high values (e.g., 3000 W) are
highly variable, since they depend on the appliance usage by the user, e.g., the water
temperature chosen in the washing machine or the rinsing cycle of the dishwasher
affects the maximum power consumption. On the other hand, clusters characterized
by low power value (e.g., 300 W) have lower variability, since deviation from the
centroid is mainly caused by intermediate working stages of the appliance, and they
do not depend on the usage.

Figure 4.8 shows an example of the inference procedure conducted on the
active power signal only, denoted as P_a, and on the joint active–reactive power
signals, denoted as (P_a, P_r). The signals are related to the washing machine in
the AMPds dataset. In particular, Fig. 4.8a shows the active power signal and the
cluster membership of each sample when k-means operates on the P_a signal only.
Figure 4.8b, c show, respectively, the same active power signal and the reactive
power signal, but the cluster membership is related to the outcome of k-means
operating on the joint (P_a, P_r). Figure 4.8d shows at the bottom the 1-D P_a line
with the clusters obtained when k-means operates on the P_a signal only and at the
top the (P_a, P_r) plane with the clusters obtained when k-means operates on the joint
(P_a, P_r) signals. In the figure, each cluster is depicted as an interval or as an ellipse
whose size is twice the standard deviation of the cluster centred at its centroid. The
number of clusters is different between the active power and the active and reactive
power cases: in the first case 4 clusters can be identified, whereas the addition of the
reactive power allows to distinguish 5 clusters. As shown in the figure, 2 clusters
share the same value of active power, but differ in the reactive component. Using
the reactive power, thus, allows to have a better representation of the working states
of the appliance, therefore reducing the admissible combination of working states
in the aggregated data.

Since the pause interval between two footprint is not recorded, the user has to
establish the time interval between two appliance activations, e.g., the typical time
of use in the daytime or the number of activations per day of the appliance, in order
to calculate the OFF interval and to use this value for the calculation of the transition
probability related to the OFF state.

4.1.2   Rest-of-the-World Model

In a real case scenario, a noise contribution can be observed on the aggregated
signal, due to electrical noises in the system, very small loads, leakages. This
contribution can be considered as a source of power consumption, additionally to
the appliances which the system tries to disaggregate, therefore it can be modelled

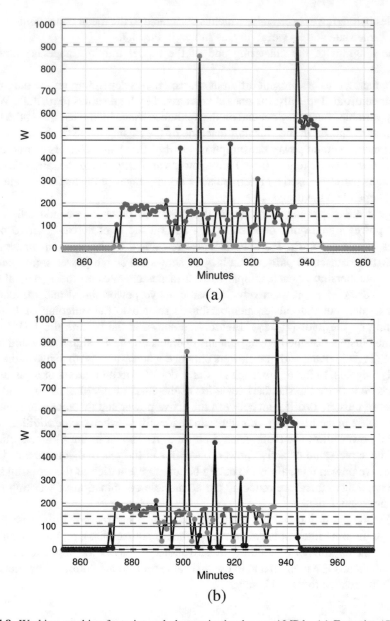

Fig. 4.8 Washing machine footprint and clusters in the dataset AMPds. (**a**) Footprint (P_a) and cluster membership of each sample with k-means operating on P_a. (**b**) Footprint (P_a) and related clusters with k-means operating on (P_a, P_r). (**c**) Footprint (P_r) and related clusters with k-means operating on (P_a, P_r). (**d**) Clusters in the (P_a, P_r) plane (above) and the P_a line (below)

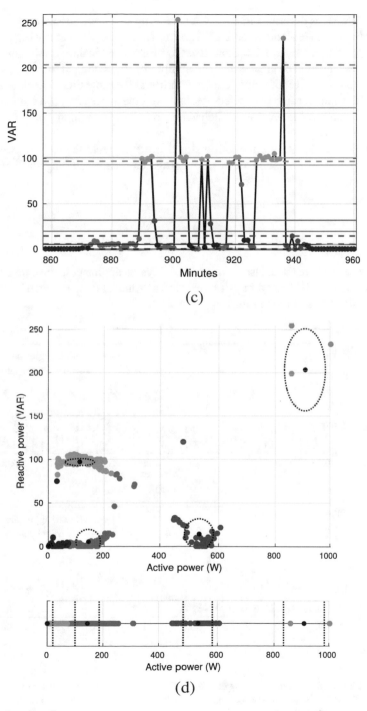

Fig. 4.8 (continued)

with an HMM, as described in Sect. 4.1.1, leading to a *noise* model or *Rest-of-the-World* (RoW) model. The number of working states is a parameter which depends on the application scenario, therefore it has to be explored in the experimental phase, nevertheless it would be greater than the number of states defined for the appliances, since it represents a set of multiple load working at the same time. The data used for training this model can be extracted by observing the aggregate power signal, when all the appliances of interest are switched off and all the remaining equipment in the house are working.

Referring to Eq. (2.1), the training signal used to create the RoW model is the residual power consumption from the aggregated data, excluding the appliances power consumption:

$$e(t) = \overline{y}(t) - \sum_{i=1}^{N} y^{(i)}(t). \tag{4.10}$$

In the case where the dataset comprises always-on appliances, since no operating cycle or footprint is defined in this case the RoW model does not include the OFF working state, as showed in Fig. 4.9.

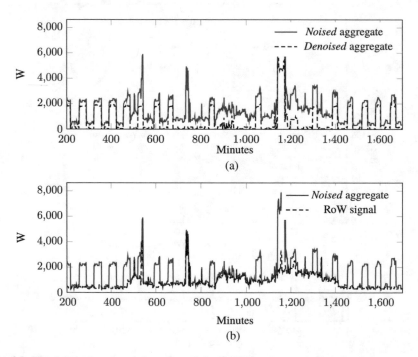

Fig. 4.9 The denoised aggregated power and the RoW signal, compared to the main aggregated power, in the AMPds. (**a**) Noised aggregated power vs denoised aggregated power. (**b**) Noised aggregated power vs RoW signal

The consumption values in the working states of the RoW model are extracted algorithmically using the k-means, even if there are no evident consumption values clusters, determined by any working state.

4.2   Algorithm Improvements

In the reference approach, the DFHMMs are obtained as the difference, in terms of power consumption, between the current and the previous sample (referred to as *backward transition*), so that a change in the state of an HMM can be evaluated against the change in the aggregated power consumption. Similarly, an additional evaluation, based on the next against the current sample (referred to as *forward transition*), is carried out. Furthermore, a smarter employment of the solver boundaries is evaluated, starting from a more accurate analysis of the aggregated power or using heterogeneous information, as the reactive power consumption of the electrical system.

Since the AFAMAP algorithm operates offline, it is possible to further extend the model by taking into account the transition from the current to the next state. The original DFHMM [21] is computed by looking backward from the current sample to the previous one, and thus it can be addressed to as Backward DFHMM. The new differential FHMM is computed by looking forward, as showed in Fig. 4.10, and thus is referred to as Forward FHMM.

The formulation of the new model, also, differs from the original one, only in the index order. The new variables define the problem, as follows:

$$\mathcal{Q} = \left\{ \boldsymbol{Q}(x^{(i)}(t)) \in \mathbb{R}^{m_i}, \ \boldsymbol{Q}(x^{(i)}(t+1), x^{(i)}(t)) \in \mathbb{R}^{m_i \times m_i} \right\},$$

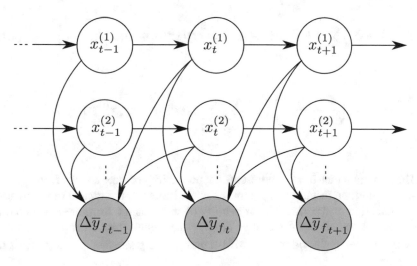

Fig. 4.10 The forward differential FHMM

where the variables are indicators of the transition from the next to the current state: $Q(x^{(i)}(t))_j = 1 \Leftrightarrow x^{(i)}(t) = S_j$, and $Q(x^{(i)}(t+1), x^{(i)}(t))_{j,k} = 1 \Leftrightarrow x^{(i)}(t+1) = S_j, x^{(i)}(t) = S_k$. The consistent constraints between the state variables and transition variables need to be satisfied:

$$
\mathcal{L} = \left\{ Q : \begin{array}{c}
\sum_{j=1}^{m_i} Q(x^{(i)}(t))_j = 1 \\
\sum_{k=1}^{m_i} Q(x^{(i)}(t+1), x^{(i)}(t))_{j,k} = Q(x^{(i)}(t+1))_j \\
\sum_{k=1}^{m_i} Q(x^{(i)}(t+1), x^{(i)}(t))_{k,j} = Q(x^{(i)}(t))_j \\
0 \le Q(x^{(i)}(t))_j, Q(x^{(i)}(t+1), x^{(i)}(t))_{k,j} \le 1
\end{array} \right\}. \tag{4.11}
$$

Therefore, the new cost function is derived from the Forward DFHMM, based on the forward differential aggregated signal $\Delta \bar{y}_f(t) = \bar{y}(t) - \bar{y}(t+1)$, as follows:

$$
\frac{1}{2\sigma_3^2} \sum_{t=1}^{T-1} E''''(t) + \frac{1}{2} \sum_{t=1}^{T-1} E'''''(t)
$$

$$
+ \sum_{t=1}^{T-1} \sum_{i=1}^{N} \sum_{\substack{j=1 \\ k=1}}^{m_i} \left\{ Q(x^{(i)}(t+1), x^{(i)}(t))_{j,k} \left(-\log P_{f_{k,j}}^{(i)} \right) \right\} \tag{4.12}
$$

$$
+ \sum_{i=1}^{N} \sum_{j=1}^{m_i} \left\{ Q(x^{(i)}(T))_j (-\log \phi_{f_j}^{(i)}) \right\},
$$

where the error terms in (4.12) are defined as:

$$
E''''(t) = \sum_{i=1}^{N} \sum_{\substack{j=1 \\ k=1 \\ k \neq j}}^{m_i} \left\{ \left(\Delta \bar{y}_f(t) - \Delta \mu_{k,j}^{(i)} \right)^2 Q(x^{(i)}(t+1), x^{(i)}(t))_{j,k} \right\}, \tag{4.13}
$$

$$
E'''''(t) = D \left(\frac{\Delta \bar{y}_f(t)}{\sigma_3}, \lambda \right) \left(1 - \sum_{i=1}^{N} \sum_{\substack{j=1 \\ k=1 \\ k \neq j}}^{m_i} Q(x^{(i)}(t+1), x^{(i)}(t))_{j,k} \right). \tag{4.14}
$$

The transition matrix $\boldsymbol{P}_f^{(i)}$ represents the probability of state change from the next to the current time instant: this parameter is equivalent to the typical representation of the transition matrix (i.e., the probability of state change from the previous time instant to the actual) evaluated after flipping the signal, thus it can be derived by using the available algorithm for HMM training. The parameter $\boldsymbol{\phi}_f^{(i)}$ represents the final state distribution, that is the initial state distribution starting from the end of the signal.

Since the duality in the forward and backward representation of the AFHMM (i.e., it is derived from the same observed signal, but in opposite directions), the problem definition using only one of the two versions of the DFHMM leads to the already known performance. Considering simultaneously both versions of DFHMM may lead to performance improvements: for this reason the forward differential function (4.12) is added to the reference formulation (4.25), thus leading to a new optimization problem.

The variable vector v in the QP problem accounts for the new terms, following the same structure introduced in Sect. 4.1:

$$\boldsymbol{\psi}^{(i)}(t) = \begin{bmatrix} \boldsymbol{\xi}^{(i)}(t) \\ \boldsymbol{\beta}^{(i)}(t) \\ \boldsymbol{\phi}^{(i)}(t) \end{bmatrix}, \qquad \boldsymbol{\phi}^{(i)}(t) = \begin{bmatrix} Q(x^{(i)}(t+1), x^{(i)}(t))_{1,1} \\ \vdots \\ Q(x^{(i)}(t+1), x^{(i)}(t))_{1,m_i} \\ \vdots \\ Q(x^{(i)}(t+1), x^{(i)}(t))_{m_i,1} \\ \vdots \\ Q(x^{(i)}(t+1), x^{(i)}(t))_{m_i,m_i} \end{bmatrix},$$

where the new term $\boldsymbol{\phi}^{(i)}(t)$ represents the variables for the forward transition.

The introduction of the new variables leads to an alteration of the problem constraints, represented by the parameters A_{eq} and b_{eq}, and the variable boundaries lb and ub. In A_{eq} the constraint about $Q(x^{(i)}(t+1), x^{(i)}(t))$ with $t = T$ has to be removed since there is no information about $Q(x^{(i)}(t))$ at the following time instant, thus falling back to the constraint $0 \cdot Q(x^{(i)}(t+1), x^{(i)}(t)) = 0$.

In order to solve the optimization problem, different solutions, which satisfy the constraints, need to be evaluated before the solver finds the optimal one. As such, the values of v that are not compatible with the given set of samples can be discarded, to restrict the search domain and improve the search efficiency.

On purpose, the lower and upper boundaries of the variable v are selected beforehand in order to prevent that the solver investigates those combinations of states that do not match the value of the aggregated power consumption. The selection method is similar to the one proposed in [89].

If several runs of a single appliance are evaluated, although the same working states are identified, the signature tends to differ from a run to the other. For this reason, the appliance power consumption can be modelled as a stochastic process and, therefore, the output value $y^{(i)}(t)$, relative to a working state $x^{(i)}(t)$ of an appliance, can be modelled as a gaussian variable, described by a mean value and a variance value:

$$y^{(i)}(t)|x^{(i)}(t) \sim \mathcal{N}\left(\mu_{x^{(i)}(t)}^{(i)}, \sigma_{x^{(i)}(t)}^{(i)\,2}\right). \tag{4.15}$$

In regard to this, the power signal is replaced by a simplified model that presents a constant power consumption, corresponding to the mean value of the working state power value, with a superimposed noisy contribution, described by the variance value in the working state.

Since the aggregated data $\overline{y}(t)$ is assumed to correspond with the sum of the power consumption of each appliance, it can be modelled as a gaussian variable, described by a mean value and a variance value equivalent to the sum of the corresponding values of each appliance, under the assumption of statistical independence between the appliances:

$$\overline{y}(t)|x^{(1:N)}(t) \sim \mathcal{N}\left(\sum_{i=1}^{N}\mu_{x^{(i)}(t)}^{(i)}, \sum_{i=1}^{N}\sigma_{x^{(i)}(t)}^{(i)\,2}\right). \tag{4.16}$$

This simplified model results in a number of admissible combinations of working states equal to $\prod_{i=1}^{N} m_i$. It allows to evaluate which combination of working states fits the power value for each sample of the aggregated data, thus discarding the incompatible ones. The effectiveness interval for each combination is centred in mean value, and its width is twice the value of the standard deviation. For some combinations, which have similar mean value or great variance, the effectiveness intervals result overlapped: for those cases, if the power value falls in this region, both the combinations are considered valid.

Based on this observation, it is possible to manipulate the boundaries of the QP problem domain. For instance, if two HMMs are considered, $M1$ and $M2$, whose power levels are $M1 = \{70, 0\}$ and $M2 = \{100, 20, 0\}$, respectively, the different combined power levels are $\{0, 20, 70, 90, 100, 170\}$, each one with its own variance value. This example is represented in Fig. 4.11. Considering a few different values of aggregated power, e.g., $\overline{y}(t) = \{20, 95, 140\}$, it can be observed that $\overline{y}(t) = 20$ is obtained as the combination $(x^{(1)}(t) = S_2, x^{(2)}(t) = S_2)$, therefore the allowed constraints are defined as:

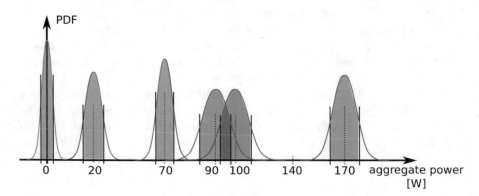

Fig. 4.11 A sketch of the different probability density functions (PDF) for each aggregated power value produced by the combination of all appliances states power levels

$$\begin{bmatrix} 0 \\ 1 \end{bmatrix} \leq \boldsymbol{\xi}^{(1)}(t) \leq \begin{bmatrix} 0 \\ 1 \end{bmatrix}, \begin{bmatrix} 0 \\ 1 \\ 0 \end{bmatrix} \leq \boldsymbol{\xi}^{(2)}(t) \leq \begin{bmatrix} 0 \\ 1 \\ 0 \end{bmatrix}.$$

If $\bar{y}(t) = 95$, the value falls in an overlapped interval, belonging to the combinations $(x^{(1)}(t) = S_2, x^{(2)}(t) = S_1)$ and $(x^{(1)}(t) = S_1, x^{(2)}(t) = S_2)$, thus, the allowed constraints are defined as:

$$\begin{bmatrix} 0 \\ 0 \end{bmatrix} \leq \boldsymbol{\xi}^{(1)}(t) \leq \begin{bmatrix} 1 \\ 1 \end{bmatrix}, \begin{bmatrix} 0 \\ 0 \\ 0 \end{bmatrix} \leq \boldsymbol{\xi}^{(2)}(t) \leq \begin{bmatrix} 1 \\ 1 \\ 0 \end{bmatrix},$$

whereas if $\bar{y}(t) = 140$, no combination is corresponding, thus the boundaries remain as default.

Clearly, the same process can be applied to bound the $\boldsymbol{\beta}^{(i)}(t)$ and $\boldsymbol{\phi}^{(i)}(t)$. In regard to this, however, since transitions are related to the steady states, the evaluation of the steady states is enough to bound both kinds of variables.

Even though disaggregation is aimed for the aggregated power consumption, in most cases the focus is centred on the active power alone. Nonetheless, given the generality of the AFAMAP algorithm, targeting the reactive aggregated power is also possible. In regard to this, in the present work, the application of the AFAMAP algorithm to the aggregated reactive power has been investigated as well, based on the fact that reactive power is a common trait of the power signature of a residential appliances subset.

In the current scenario, the disaggregation of the reactive power samples is carried out, in order to collect additional information about the activity states of the appliances. This information, in turn, is used to further define the lower and the upper boundaries of the states in the active power disaggregation. Similarly to the active power case, the HMMs are modelled for each appliances starting from the signature in the reactive power and the AFAMAP algorithm is run by using the aggregated reactive power signal as input.

Following the basic knowledge in circuit theory, an electrical load with a reactive component (i.e., an appliance) which has a reactive power consumption greater than 0 is necessarily turned on, therefore the boundaries of the problem in active power disaggregation are assigned as follows:

$$\begin{bmatrix} 0 \\ 0 \\ 0 \end{bmatrix} \leq \boldsymbol{\xi}^{(i)}(t) \leq \begin{bmatrix} 1 \\ 1 \\ 0 \end{bmatrix}.$$

Although when the reactive power consumption is 0, the active component could be both null or greater than 0, depending on whether the appliance is turned off or only the load passive component is working. Therefore, the boundaries of the problem in active power disaggregation are set as default.

4.2.1   Experimental Setup

The dataset used for the experiments is the Almanac of Minutely Power dataset (AMPds) [58]: it contains recordings of consumption profiles belonging to a single home in Canada for a period of 2 years at 1 min sampling rate. It provides active and reactive power at appliance level, unlike most of the dataset in which only the active power is provided at appliance level, as described in Sect. 2.3: this information is crucial to test the new approach based on the reactive power disaggregation as constraint. Analysing the contents of the dataset, it can be noticed that the usage of the appliances is homogeneous throughout the entire period, therefore the experiments are evaluated on 6 months of data, which can be considered a representative of the entire dataset. To create the HMM models of the appliances, the training requires at least one signature per appliance, although multiple signatures lead to a more general model. In the proposed work, a subset of the data, spanning over 14 days, has been deemed sufficient to collect all the signatures required to train all the HMMs. The HMM are trained in accordance with the Baum-Welch algorithm, after determining the ground truth state over the time: those are obtained through a clustering procedure, in which every cluster represents a power consumption level of the appliance, thus a state of the HMM. This process is achieved using the k-means algorithm, in which the number of the clusters is imposed in a supervised manner, starting from the knowledge of the operating states of the appliance. The power level mean and the variance values are achieved by means of a gaussian variable fitting procedure over the samples belonging to each cluster. To satisfy the condition of *denoised* system, the aggregated data is synthetically composed by summing the appliance level power signals. The experiments are conducted by using the appliances at higher contribution, therefore 6 appliances have been chosen: dryer, washing machine, dishwasher, fridge, oven and heat pump. The simulations are conducted in Matlab environment and the CPLEX solver is used to solve the QP problem. The value of starting probability $\phi_b^{(i)}$ of the i-th HMM is imposed to assume the certainty for the OFF state for $f = 1$, whereas for the consecutive windows, $1 < f \leq F$, it is imposed to assume the value of the last sample $\xi^{(i)}(T)$ of the previous window, in order to ensure the contiguity of the solution on the window border. The value of the ending probability $\phi_f^{(i)}$, instead, is uniformly imposed in every state, since no information from the consecutive window is available. Different experiments are conducted varying the size of the windows between the values $T \in \{10, 30, 60, 90, 120\}$ min, and the effectiveness of the innovative aspect is evaluated: the introduction of the forward term in the cost function, the selection of the boundaries related to the aggregated power level and to the disaggregation output of the reactive power. The variance parameters are defined with $\sigma_1^2 = \sigma_2^2 = \sigma_3^2 = 0.01$ according to the variance of the experimental data and the regularization parameter $\lambda = 1$.

4.2.2   Results

The results of the experiments, based on the scenario described in Sect. 4.2.1, are presented in the current section.

In Fig. 4.12, the AFAMAP disaggregated power consumption profiles of the appliances are compared against the corresponding true outputs, provided by the dataset: in the figure a time span of 10 h, corresponding to 600 samples, is considered. At the bottom, the energy distribution over the same period, expressed among

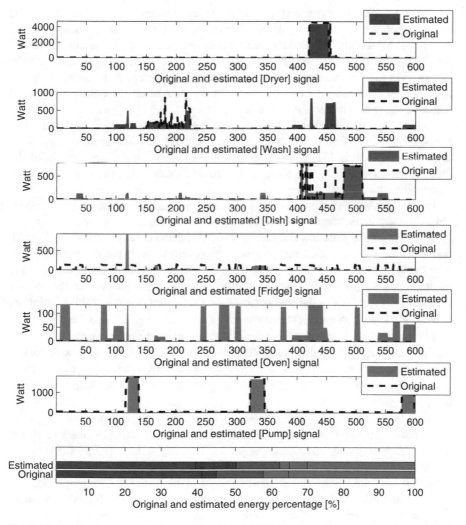

Fig. 4.12 Appliances consumption: estimated AFAMAP disaggregation output against original signals

the appliances in terms of percent value, is compared between the reconstructed and the true appliances consumption.

The signals reveal that the appliances which show a high steady power consumption are easily recognized, whereas the appliances with complex working cycles, or with several power levels, are more difficult to detect. Indeed, whenever several appliances present similar consumption levels, many combinations may satisfy the problem constraints, thus additional information is required to identify the active appliances. For instance, in Fig. 4.12, the oven and the fridge are seldom recognized, whereas the detection of the dryer and the washing machine are partially more successful.

The evaluation of the algorithm performance is carried out by means of the metrics proposed in Sect. 2.4. Although the focus of the present work is on the AFAMAP algorithm, the dataset being used and the proposed training method are different with respect to [21], therefore a direct comparison against the results proposed in the reference work is not possible. To overcome this shortcoming, the baseline has been created anew, by means of the AFAMAP algorithm, the AMPds dataset and the proposed training method.

The disaggregation results computed by means of the metrics are reported in Fig. 4.13: in Fig. 4.13a the state based metric is presented, whereas the energy based metric is proposed in Fig. 4.13b. The results are shown for different values of the time window length. Clearly, since all the results exceed 0.5, the plots have been drawn from 0.5 onwards.

Both plots show that the best results are achieved using the shortest time window. On a different note, however, not every configuration improves in the same way.

Focusing on the state based metrics, it is possible to observe that the AFAMAP baseline shows a significant performance improvement with the decreasing of the window length, except when passing from the 30 to 10 min window size. On the contrary, the forward differential model shows an improvements at the shorter window size, resulting in the best performance in the unbounded problem solution, with an $F_1^{(S)}$ of 0.738 and an improvement of 1% with respect to the baseline.

Fixing the boundaries of the problem, in every simulation case, gives the benefit on the disaggregation results: the profile based method gives a considerable performance improvements with every window size, but the highest relative improvement can be noted at the smallest size, resulting to an $F_1^{(S)}$ of 0.863 and a relative improvement of 18%.

Alternatively, the boundaries can be set based on the reactive power disaggregation feedback: the results, showed in Table 4.2, demonstrate that the reactive power reaches high performance in disaggregation. This is due to the high difference in the reactive components of each appliance, which involves a strong distinction in the creation of the HMM, therefore allowing a highly reliable disaggregation. The usage of this information results in a performance improvement for every window size, more considerable at the smallest size: in general, the usage of the reactive power feedback gives benefits to the disaggregation, with an $F_1^{(S)}$ of 0.802 and a relative improvement of 9.7%, therefore less than the profile based constraints.

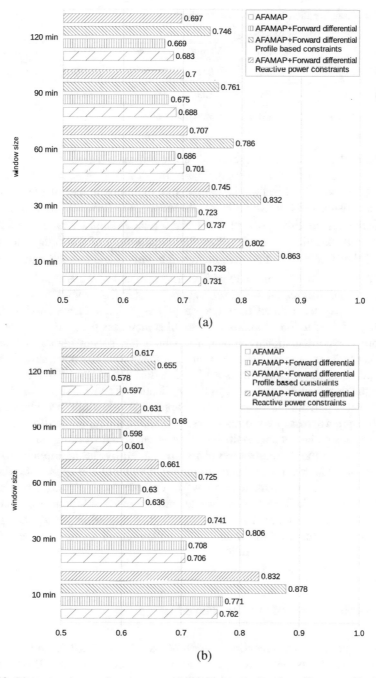

Fig. 4.13 Disaggregation performance on AMPds dataset using 6 appliances, with different algorithm configuration. (**a**) State based metric: $F_1^{(S)}$. (**b**) Energy based metric: $F_1^{(E)}$

Table 4.2 Disaggregation results on reactive power

Metric	Window size				
	10 min	30 min	60 min	90 min	120 min
State based: $F_1^{(S)}$	0.922	0.877	0.869	0.867	0.865
Energy based: $F_1^{(E)}$	0.935	0.883	0.877	0.875	0.874

The configuration used is: AFAMAP + Forward differential

Clearly, the same trends presented about the state based metrics still hold true when evaluating with the energy based metrics. The most notably difference between the two plots, in fact, is that the rate of improvement of the algorithms when decreasing the time window length: indeed, the forward differential model introduction results to an $F_1^{(E)}$ of 0.771 and a relative improvement of 1.2% with respect to the baseline, whereas the profile based setting of the boundaries results to an $F_1^{(E)}$ of 0.878 with a relative improvement of 15.2% and the reactive power based method to an $F_1^{(E)}$ of 0.832 with an improvement of 9.2%.

The forward differential model seems to be beneficial only with the shortest time window: it may be a direct consequence of the problem formulation alteration. Indeed, the introduction of additional variables increases the size of the problem, therefore the computational burden, for which the solver demonstrates worst performance, as it happens for the baseline approach with larger window size.

Despite this, the improvements achieved adding the differential forward information to the model are restricted to the application scenario: since the algorithm operates on a per-sample basis, for each appliance behaviour two state changes unlikely happen across three contiguous samples, thus the forward difference cannot provide a substantial support to the inference of the actual working state.

The errors in the disaggregation phase are caused by the multiplicity of states combinations which can correspond to the same value of the aggregated data: for this reason the use of boundaries allows to exclude some solutions that are not eligible, therefore facilitates the solver to find the exact solution to the problem. Nevertheless, the variation over time of the power consumption associated to a specific appliance working state causes an unwanted variability, i.e., a noise component, in the achieved solution.

4.3   Exploitation of the Reactive Power

In this section, a disaggregation algorithm based on FHMMs and active and reactive power measured at low sampling rates is proposed. The HMM models of the appliances and the proposed solution for obtaining their parameters from a training dataset are described. Load disaggregation is performed by proposing a reformulated version of the Additive Factorial Approximate Maximum a Posteriori

(AFAMAP) algorithm [21] that allows a straightforward extension to the bivariate case. The experimental evaluation has been conducted on the Almanac of Minutely Power dataset (AMPds) dataset [58] in noised and denoised scenarios, and the proposed solution has been compared to AFAMAP based on the active power only and to two variants of Hart's algorithm [15] both based on active and reactive power. The results show that in terms of energy based F_1-Measure ($F_1^{(E)}$) the proposed approach provides a significant performance improvement with respect to the comparative methods.

Apart from [20], the aforementioned approaches employ the active power as the sole electrical parameter for NILM, despite some algorithmic frameworks have been formulated for operating on multidimensional feature vectors [21]. The reactive power has been employed since the very first work by Hart [15] and in more recent works based on the same principles [90–94] or on transient-state analysis [46–50]. However, to the best of author's knowledge, the only work that employs both the active and reactive power in the FHMM framework is the work by Zoha and colleagues [20].

Following a similar philosophy, a disaggregation algorithm based on FHMMs that uses both the active and reactive power is proposed. However, differently from [20], where the disaggregation algorithm is based on the structural variational approximation method and on the Viterbi algorithm, in the proposed approach the active power is disaggregated by reformulating the AFAMAP algorithm for the bivariate case. As demonstrated in [21], this allows the introduction of a Differential FHMM (DFHMM) that improves the performance and reduces the computational cost. Thus, differently from [20], here the reactive power component is introduced also in the DFHMM. More in details, the proposed solution belongs to the family of supervised approaches based on steady-state signals acquired from low frequency measurements. The reactive power is introduced in the FHMM framework by employing bivariate hidden Markov appliance models whose emitted symbols are represented by active and reactive power pairs. Differently from [20], the entire procedure for obtaining the bivariate HMM appliance models is described. The parameters are estimated by clustering the appliance disaggregate signals and the bivariate optimization problem is solved by proposing an alternative formulation of AFAMAP [21] for disaggregating appliances consumption profiles. The proposed approach differs from the one presented in Sect. 4.2, since there the reactive power was employed alone in an initial disaggregation stage whose output served as a constraint for the subsequent disaggregation of the active power only. The proposed approach has been compared to the original AFAMAP algorithm [21], which employs the active power only, and to Hart's algorithm [15], which employs both the active and reactive power. In order to deal with the occurence of multiple appliance combination, two implementation of Hart's algorithm have been developed: in the first, the final combination is selected randomly. In the second, it is selected by choosing the most probable combination calculated on a training set. The experiments have been conducted on the Almanac of Minutely Power dataset (AMPds) [58], containing recordings of consumption profiles belonging to a single home for a period of 2 years at 1 min sampling rate. Both the "noised" and the "denoised" scenarios have been addressed, and the results show that the proposed approach outperforms both AFAMAP and Hart's algorithm.

Finally, in [20] the experiments are conducted on low-power appliances only in a "denoised scenario", while here the "noised" is also considered.

In the following, the superscript (i) denotes terms related to HMM i, while subscripts a or r denote terms related to the active and reactive power components, respectively. The subscript $c \in \{a, r\}$ denotes a term related to the active or to the reactive power component. The parameters of the problem are the following:

- $N \in \mathbb{Z}_+$ is the number of HMMs in the system;
- $\overline{y}(\tau) \in \mathbb{R}^n$ is the observed aggregate output, where $\tau = 1, 2, \ldots, \Upsilon$ is the sample index and Υ is the total number of samples;
- $\Sigma_1 \in \mathbb{R}^{n \times n}$ is the observation covariance matrix related to the AFHMM;
- $\Sigma_2 \in \mathbb{R}^{n \times n}$ is the observation covariance matrix related to the DFHMM;
- $\Delta \overline{y}(\tau) = \overline{y}(\tau) - \overline{y}(\tau - 1)$ is the differential signal.

As aforementioned, all the contribution to the aggregated power are considered, thus:

$$\overline{y}(\tau) = \sum_{i=1}^{N} y^{(i)}(\tau), \tag{4.17}$$

where $y^{(i)}(\tau)$ corresponds to the ground truth consumption of the appliances and the *noise*. Recalling the notation of Chap. 4, the parameters of the i-th HMM at the sample index τ are:

- $m_i \in \mathbb{Z}_+$ is the number of states;
- $x^{(i)}(\tau) \in \{S_1, \ldots, S_{m_i}\}$ is the HMM state at time instant τ, where S_{m_i} corresponds to the OFF state (if present);
- $\mu_j^{(i)}$ is the emitted symbol in the j-th state, where $j = 1, 2, \ldots, m_i$;
- $\phi^{(i)} \in [0, 1]^{m_i}$ is the initial states probability distribution;
- $P^{(i)} \in [0, 1]^{m_i \times m_i}$ is the state transition probability matrix.

The aggregate signal $\overline{y}(\tau)$ is analysed using non-overlapping frames of length T. Each frame $\overline{y}_f(\tau)$, where $f = 1, 2, \ldots, F$, is defined as

$$\overline{y}_f(\tau) = \begin{cases} \overline{y}(\tau) & \text{if } \tau = (f-1)T + 1, \ldots, fT, \\ 0 & \text{otherwise.} \end{cases} \tag{4.18}$$

After the analysis of all the $F = \Upsilon/T$ frames, the disaggregated signals $\hat{y}^{(i)}(\tau)$ are reconstructed as follows:

$$\hat{y}^{(i)}(\tau) = \sum_{f=1}^{F} \hat{y}_f^{(i)}(\tau). \tag{4.19}$$

In the following, the algorithm is formulated for a single frame of the signal and for convenience, a new temporal variable t is defined with the relation $t = \tau - (f-1)T$, for $t = 1, 2, \ldots, T$, with $T \in \mathbb{Z}_+$.

In [21], the parameter n defines the problem dimensionality: the authors use only the active power data to characterize the observed aggregated signal $\overline{y}_a(t)$, therefore they assumed $n = 1$. In this work, both the active and the reactive power are used for disaggregation, therefore $n = 2$ and the problem variables are decomposed in two components:

$$\overline{y}_f(t) = \begin{bmatrix} \overline{y}_{a,f}(t) \\ \overline{y}_{r,f}(t) \end{bmatrix}, \quad \mu_j^{(i)} = \begin{bmatrix} \mu_{a,j}^{(i)} \\ \mu_{r,j}^{(i)} \end{bmatrix}, \tag{4.20}$$

$$\Sigma_1 = \begin{bmatrix} \sigma_{a,1}^2 & \sigma_{a,r,1} \\ \sigma_{r,a,1} & \sigma_{r,1}^2 \end{bmatrix}, \quad \Sigma_2 = \begin{bmatrix} \sigma_{a,2}^2 & \sigma_{a,r,2} \\ \sigma_{r,a,2} & \sigma_{r,2}^2 \end{bmatrix}. \tag{4.21}$$

Since the statistical independence between the active and reactive power components is supposed, the covariance terms $\sigma_{a,r}$ and $\sigma_{r,a}$ are zero in both Σ_1 and Σ_2, and the same problem formalization as the $n = 1$ case can be used, introducing additional variables and constraining them each other. For the generic power component c, the variables in the optimization problem are defined as follows:

$$\mathcal{Q}_c = \left\{ \boldsymbol{Q}_c(x^{(i)}(t)) \in \mathbb{R}^{m_i}, \boldsymbol{Q}_c(x^{(i)}(t-1), x^{(i)}(t)) \in \mathbb{R}^{m_i \times m_i} \right\}. \tag{4.22}$$

In the vector $\boldsymbol{Q}_c(x^{(i)}(t))$, the element $Q_c(x^{(i)}(t))_j$ indicates the state assumed at time instant t, while in the matrix $\boldsymbol{Q}_c(x^{(i)}(t-1), x^{(i)}(t))$ the element $Q_c(x^{(i)}(t-1), x^{(i)}(t))_{jk}$ indicates the state transition from previous to the current time instant.

This problem statement is a reformulated version of the algorithm proposed in [21]: since the original algorithm allows to operate with multivariate dimension, the variables associated to the state represent all the components. When only one dimension is considered, the variables $\mathcal{Q}_a$ is only associated at the active power level consumption. This problem statement instead started from the univariate formulation, and the algorithm is extended to $n = 2$ by using twice the optimization variables, thus introducing the $\mathcal{Q}_r$ variable set, and an additional minimization function. Moreover, the supplementary variables need to be constrained to the original ones in order to assume the same value during the optimization process, representing the bivariate resolution problem with a univariate problem formalization:

$$\begin{cases} Q_a(x^{(i)}(t))_j - Q_r(x^{(i)}(t))_j = 0, \\ Q_a(x^{(i)}(t-1), x^{(i)}(t))_{jk} - Q_r(x^{(i)}(t-1), x^{(i)}(t))_{jk} = 0. \end{cases} \tag{4.23}$$

A numerically safer definition of the constraints can be defined using a tolerance α and inequalities:

$$\begin{cases} -\alpha \le Q_a(x^{(i)}(t))_j - Q_r(x^{(i)}(t))_j \le \alpha, \\ -\alpha \le Q_a(x^{(i)}(t-1), x^{(i)}(t))_{jk} - Q_r(x^{(i)}(t-1), x^{(i)}(t))_{jk} \le \alpha, \end{cases} \qquad (4.24)$$

where $j, k = 1, \ldots, m_i$.

Algorithm 1 The proposed disaggregation algorithm
___
1: **Input**:

- $\overline{y}_f(t)$, for $t = 1, 2, \ldots, T$;
- $\{\mu^{(i)}, P^{(i)}, \phi^{(i)}\}$, for $i = 1, 2, \ldots, N$;
- $\sigma_{c,1}^2, \sigma_{c,2}^2$;
- λ: regularisation parameter, described in [21].

2: **Minimise over** $\{Q_c \in \mathcal{L}_c \cap \mathcal{O}_c\}$

$$\sum_{c \in \{a,r\}} \left\{ \frac{1}{2\sigma_{c,1}^2} \sum_{t=1}^{T} E_c'(t) + \frac{1}{2\sigma_{c,2}^2} \sum_{t=2}^{T} E_c''(t) + \frac{1}{2} \sum_{t=2}^{T} E_c'''(t) + \right.$$

$$+ \sum_{t=2}^{T} \sum_{i=1}^{N} \sum_{\substack{j=1 \\ k=1}}^{m_i} \left\{ Q_c(x^{(i)}(t-1), x^{(i)}(t))_{jk} \left(-\log P_{kj}^{(i)} \right) \right\} +$$

$$\left. + \sum_{i=1}^{N} \sum_{j=1}^{m_i} \left\{ Q_c(x^{(i)}(1))_j (-\log \phi_j^{(i)}) \right\} \right\} \qquad (4.25)$$

3: **Output**:

$$\hat{y}_{c,f}^{(i)}(t) = \sum_{j=1}^{m_i} \mu_{c,j}^{(i)} Q_c(x^{(i)}(t))_j \qquad (4.26)$$

where $i = 1, 2, \ldots, N$ and $t = 1, 2, \ldots, T$.
___

The final algorithm is shown in Algorithm 1. In Eq. (4.25), the error terms are defined as:

$$E_c'(t) = \left(\overline{y}_{c,f}(t) - \sum_{i=1}^{N} \sum_{j=1}^{m_i} \mu_{c,j}^{(i)} Q_c(x^{(i)}(t))_j \right)^2, \qquad (4.27)$$

$$E_c''(t) = \sum_{i=1}^{N} \sum_{\substack{j=1 \\ k=1 \\ k \ne j}}^{m_i} \left\{ \left(\Delta \overline{y}_{c,f}(t) - \Delta \mu_{c,kj}^{(i)} \right)^2 Q_c(x^{(i)}(t-1), x^{(i)}(t))_{jk} \right\}, \qquad (4.28)$$

$$E_c'''(t) = D\left(\frac{\Delta\bar{y}_{c,f}(t)}{\sigma_{c,2}}, \lambda\right)\left(1 - \sum_{i=1}^{N}\sum_{\substack{j-1 \\ k=1 \\ k\neq j}}^{m_i} Q_c(x^{(i)}(t-1), x^{(i)}(t))_{jk}\right). \quad (4.29)$$

The QP optimization problem is defined as follows:

Minimize

$$\frac{1}{2}v^T Hv + f^T v, \quad (4.30)$$

subject to the constraints:

$$A_{eq}\,v = b_{eq}, \quad (4.31)$$

$$lb \le v \le ub. \quad (4.32)$$

The variables of the problem are represented by the vector $v = [v_a\,v_r]^T$ whose components are defined as follows:

$$v_c = \begin{bmatrix} \Theta(1) \\ \vdots \\ \Theta(T) \end{bmatrix}, \quad \Theta(t) = \begin{bmatrix} \Psi^{(1)}(t) \\ \vdots \\ \Psi^{(N)}(t) \end{bmatrix}, \quad \Psi^{(i)}(t) = \begin{bmatrix} \xi^{(i)}(t) \\ \beta^{(i)}(t) \end{bmatrix}, \quad (4.33)$$

$$\xi^{(i)}(t) = \begin{bmatrix} Q_c(x^{(i)}(t))_1 \\ \vdots \\ Q_c(x^{(i)}(t))_{m_i} \end{bmatrix}, \quad \beta^{(i)}(t) = \begin{bmatrix} Q_c(x^{(i)}(t-1), x^{(i)}(t))_{11} \\ \vdots \\ Q_c(x^{(i)}(t-1), x^{(i)}(t))_{1\,m_i} \\ \vdots \\ Q_c(x^{(i)}(t-1), x^{(i)}(t))_{m_i\,1} \\ \vdots \\ Q_c(x^{(i)}(t-1), x^{(i)}(t))_{m_i\,m_i} \end{bmatrix},$$

$$(4.34)$$

where the variables for the state are represented in $\xi^{(i)}(t)$, and the variables for the transition in $\beta^{(i)}(t)$.

The parameters of the problem, e.g., the HMMs parameters and the aggregated power signal, comprise the elements of H and f, according to the structure of the v vector. In a QP problem, the coefficient of the quadratic terms in the cost function is defined in H, as a symmetric matrix. In the proposed approach, since the independence between the active and reactive power is assumed, there are no joint quadratic terms, therefore H is structured as follows:

$$H = \begin{bmatrix} H_a & 0 \\ 0 & H_r \end{bmatrix}. \tag{4.35}$$

Differently, the coefficients of the linear terms are expressed in $f = [f_a \ f_r]^T$, whereas A_{eq} and b_{eq} are used to represent the consistent constraints between the state and the transition variables. The vectors lb and ub define the lower and upper boundaries of the solution: because of the nature of the variables [21], the lower boundary is equal to 0, whereas the upper boundary to 1, for all the elements in v.

Additional constraints to QP problem need to be considered, in order to impose the inequality constraints between the optimization variables. Duplicating the constraints of Eq. (4.24):

$$\begin{cases} -\alpha \leq Q_a(x^{(i)}(t))_j - Q_r(x^{(i)}(t))_j, \\ Q_a(x^{(i)}(t))_j - Q_r(x^{(i)}(t))_j \leq \alpha, \end{cases} \tag{4.36}$$

$$\begin{cases} -\alpha \leq Q_a(x^{(i)}(t-1), x^{(i)}(t))_{jk} - Q_r(x^{(i)}(t-1), x^{(i)}(t))_{jk}, \\ Q_a(x^{(i)}(t-1), x^{(i)}(t))_{jk} - Q_r(x^{(i)}(t-1), x^{(i)}(t))_{jk} \leq \alpha, \end{cases} \tag{4.37}$$

results in the following optimization constraint:

$$A_{ineq} \, v \leq b_{ineq}. \tag{4.38}$$

This is needed only for the joint active–reactive problem, since, solving only for the active power, the related unique variable is not constrained to other variables. Indeed, in Eq. (4.25) only the active power terms need to be considered. Further details on the terms H, f, A_{eq}, b_{eq}, lb, ub, A_{ineq} and b_{ineq} are provided in 4.3.1.

As aforementioned, the aggregate signal is analysed in frames of length T. In the first frame, the value of starting probability vector $\phi^{(i)} = [0 \ 0 \ \cdots \ 0 \ 1]$, i.e., the appliance is initially assumed in the OFF state. In the subsequent frames, the value of $\phi^{(i)}$ depends on the last state assumed in the previous frame in order to ensure the contiguity of the solution at the border. Thus, if the last state assumed in the previous frame is j, the corresponding element of $\phi^{(i)}$ is set to 1, while the others are set to 0. This information is represented by the value of the solution $\xi^{(i)}(t)$ in the last sample $t = T$.

4.3.1 AFAMAP Formulation

This subsection provides further details on the algorithm formulation presented in Sect. 4.3. In particular, the following terms of the QP problem are described: H, f, A_{eq}, b_{eq}, lb, ub, A_{ineq} and b_{ineq}.
The matrix H is structured as follows:

$$H = \begin{bmatrix} H_a & 0 \\ 0 & H_r \end{bmatrix}, \tag{4.39}$$

where $H_c \in \{H_a, H_r\}$ is given by:

$$H_c = \begin{bmatrix} \frac{1}{\sigma_{c,1}^2} H_{\Theta(1)} & \cdots & 0 \\ \vdots & \ddots & \vdots \\ 0 & \cdots & \frac{1}{\sigma_{c,1}^2} H_{\Theta(T)} \end{bmatrix}, \quad H_{\Theta(t)} = \begin{bmatrix} H_{\psi(1\,1)(t)} & \cdots & H_{\psi(1\,N)(t)} \\ \vdots & \ddots & \vdots \\ H_{\psi(N\,1)(t)} & \cdots & H_{\psi(N\,N)(t)} \end{bmatrix}, \tag{4.40}$$

and

$$H_{\psi(i\,j)(t)} = \begin{bmatrix} H_{\xi(i\,j)(t)} & 0 \\ 0 & 0 \end{bmatrix}, \quad H_{\xi(i\,j)(t)} = \begin{bmatrix} \mu_{c,1}^{(i)} \mu_{c,1}^{(j)} & \cdots & \mu_{c,1}^{(i)} \mu_{c,m_j}^{(j)} \\ \vdots & \ddots & \vdots \\ \mu_{c,m_i}^{(i)} \mu_{c,1}^{(j)} & \cdots & \mu_{c,m_i}^{(i)} \mu_{c,m_j}^{(j)} \end{bmatrix}. \tag{4.41}$$

Regarding the vector f, in Sect. 4.3 it has been defined as follows:

$$f = \begin{bmatrix} f_a & f_r \end{bmatrix}^T, \tag{4.42}$$

where $f_c \in \{f_a, f_r\}$ is given by the sum of five terms:

$$f_c = -f_{c,1} - \frac{1}{\sigma_{c,1}^2} f_{c,2} - f_{c,3} + \frac{1}{2} \frac{1}{\sigma_{c,2}^2} f_{c,4} - \frac{1}{2} f_{c,5}, \tag{4.43}$$

where

$$f_{c,1} = \begin{bmatrix} f_{1,\Theta(1)} \\ 0 \\ \vdots \\ 0 \end{bmatrix}, \quad f_{1,\Theta(1)} = \begin{bmatrix} f_{1,\psi(1)(1)} \\ \vdots \\ f_{1,\psi(N)(1)} \end{bmatrix}, \quad f_{1,\psi(i)(1)} = \begin{bmatrix} f_{1,\xi(i)(1)} \\ 0 \end{bmatrix}, \quad f_{1,\xi(i)(1)} = \begin{bmatrix} \log \phi_1^{(i)} \\ \vdots \\ \log \phi_{m_i}^{(i)} \end{bmatrix}, \tag{4.44}$$

$$f_{c,2} = \begin{bmatrix} f_{2,\Theta(1)} \\ \vdots \\ f_{2,\Theta(T)} \end{bmatrix}, \quad f_{2,\Theta(t)} = \begin{bmatrix} f_{2,\psi(1)(t)} \\ \vdots \\ f_{2,\psi(N)(t)} \end{bmatrix}, \quad f_{2,\psi(i)(t)} = \begin{bmatrix} f_{2,\xi(i)(t)} \\ 0 \end{bmatrix}, \quad f_{2,\xi(i)(t)} = \begin{bmatrix} \bar{y}_{c,f}(t) \mu_{c,1}^{(i)} \\ \vdots \\ \bar{y}_{c,f}(t) \mu_{c,m_i}^{(i)} \end{bmatrix}, \tag{4.45}$$

$$f_{c,3} = \begin{bmatrix} \mathbf{0} \\ f_{3,\boldsymbol{\Theta}(2)} \\ \vdots \\ f_{3,\boldsymbol{\Theta}(T)} \end{bmatrix}, \ f_{3,\boldsymbol{\Theta}(t)} = \begin{bmatrix} f_{3,\boldsymbol{\psi}(1)(t)} \\ \vdots \\ f_{3,\boldsymbol{\psi}(N)(t)} \end{bmatrix}, \ f_{3,\boldsymbol{\psi}(i)(t)} = \begin{bmatrix} \mathbf{0} \\ f_{3,\boldsymbol{\beta}(i)(t)} \end{bmatrix}, \ f_{3,\boldsymbol{\beta}(i)(t)} = \begin{bmatrix} \log P_{11}^{(i)} \\ \vdots \\ \log P_{m_i\,1}^{(i)} \\ \vdots \\ \log P_{1\,m_i}^{(i)} \\ \vdots \\ \log P_{m_i\,m_i}^{(i)} \end{bmatrix}, \tag{4.46}$$

$$f_{c,4} = \begin{bmatrix} \mathbf{0} \\ f_{4,\boldsymbol{\Theta}(2)} \\ \vdots \\ f_{4,\boldsymbol{\Theta}(T)} \end{bmatrix}, \ f_{4,\boldsymbol{\Theta}(t)} = \begin{bmatrix} f_{4,\boldsymbol{\psi}(1)(t)} \\ \vdots \\ f_{4,\boldsymbol{\psi}(N)(t)} \end{bmatrix}, \ f_{4,\boldsymbol{\psi}(i)(t)} = \begin{bmatrix} \mathbf{0} \\ f_{4,\boldsymbol{\beta}(i)(t)} \end{bmatrix}, \tag{4.47}$$

$$f_{4,\boldsymbol{\beta}(i)(t)} = \begin{bmatrix} k_{11}^{(i)}(t) \\ \vdots \\ k_{m_i\,1}^{(i)}(t) \\ \vdots \\ k_{1\,m_i}^{(i)}(t) \\ \vdots \\ k_{m_i\,m_i}^{(i)}(t) \end{bmatrix}, \tag{4.48}$$

$$k^{(i)}(t) = \begin{bmatrix} 0 & \cdots & \left(\Delta\bar{y}_{c,f}(t) - \left(\mu_{c,1}^{(i)} - \mu_{c,m_i}^{(i)}\right)\right)^2 \\ \vdots & \ddots & \vdots \\ \left(\Delta\bar{y}_{c,f}(t) - \left(\mu_{c,m_i}^{(i)} - \mu_{c,1}^{(i)}\right)\right)^2 & \cdots & 0 \end{bmatrix} \tag{4.49}$$

$$f_{c,5} = \begin{bmatrix} \mathbf{0} \\ f_{5,\boldsymbol{\Theta}(2)} \\ \vdots \\ f_{5,\boldsymbol{\Theta}(T)} \end{bmatrix}, \ f_{5,\boldsymbol{\Theta}(t)} = \begin{bmatrix} f_{5,\boldsymbol{\psi}(1)(t)} \\ \vdots \\ f_{5,\boldsymbol{\psi}(N)(t)} \end{bmatrix}, \ f_{5,\boldsymbol{\psi}(i)(t)} = \begin{bmatrix} \mathbf{0} \\ f_{5,\boldsymbol{\beta}(i)(t)} \end{bmatrix}. \tag{4.50}$$

$$f_{5,\boldsymbol{\beta}(i)(t)} = \begin{bmatrix} d_{11}^{(i)}(t) \\ \vdots \\ d_{m_i\,1}^{(i)}(t) \\ \vdots \\ d_{1\,m_i}^{(i)}(t) \\ \vdots \\ d_{m_i\,m_i}^{(i)}(t) \end{bmatrix}, \ d^{(i)}(t) = \begin{bmatrix} 0 & \cdots & D\left(\frac{\Delta\bar{y}_{c,f}(t)}{\sigma_{c,2}}, \lambda\right) \\ \vdots & \ddots & \vdots \\ D\left(\frac{\Delta\bar{y}_{c,f}(t)}{\sigma_{c,2}}, \lambda\right) & \cdots & 0 \end{bmatrix}, \tag{4.51}$$

where:

$$D(y, \lambda) = \min\left\{\frac{1}{2}y^2, \max\left\{\lambda|y| - \frac{\lambda^2}{2}, \frac{\lambda^2}{2}\right\}\right\}. \tag{4.52}$$

The matrix A_{eq} is defined as follows:

$$A_{eq} = \begin{bmatrix} A_{eq,a} & 0 \\ 0 & A_{eq,r} \end{bmatrix}, \quad A_{eq,c} = \begin{bmatrix} A_{eq,\Theta(1)} & \cdots & 0 \\ \vdots & \ddots & \vdots \\ 0 & \cdots & A_{eq,\Theta(T)} \end{bmatrix}, \tag{4.53}$$

$$A_{eq,\Theta(t)} = \begin{bmatrix} A_{eq,\Psi_1^{(1)}(t)} & 0 & \cdots & 0 & A_{eq,\Psi_2^{(1)}(t)} & 0 & \cdots & 0 \\ \vdots & \ddots & & \vdots & & \ddots & & \vdots \\ 0 & \cdots & 0 & A_{eq,\Psi_1^{(N)}(t)} & 0 & \cdots & 0 & A_{eq,\Psi_2^{(N)}(t)} \end{bmatrix}, \tag{4.54}$$

$$A_{eq,\Psi_1^{(i)}(t)} = \begin{bmatrix} A_{eq,\xi_1^{(i)}(t)} \\ A_{eq,\beta_{1b}^{(i)}(t)} \\ A_{eq,\beta_{1f}^{(i)}(t)} \end{bmatrix}, \quad A_{eq,\xi_1^{(i)}(t)} = \begin{bmatrix} 0 & \cdots & 0 & 0 & \cdots & 0 \end{bmatrix}, \tag{4.55}$$

$$A_{eq,\beta_{1b}^{(i)}(t)} = \begin{bmatrix} -1 & \cdots & 0 & 0 & \cdots & 0 \\ \vdots & \ddots & \vdots & \vdots & \ddots & \vdots \\ 0 & \cdots & -1 & 0 & \cdots & 0 \end{bmatrix}, \quad A_{eq,\beta_{1f}^{(i)}(t)} = \begin{bmatrix} 0 & \cdots & 0 & 0 & \cdots & 0 \\ \vdots & \ddots & \vdots & \vdots & \ddots & \vdots \\ 0 & \cdots & 0 & 0 & \cdots & 0 \end{bmatrix} \tag{4.56}$$

$$A_{eq,\Psi_2^{(i)}(t)} = \begin{bmatrix} A_{eq,\xi_2^{(i)}(t)} \\ A_{eq,\beta_{2b}^{(i)}(t)} \\ A_{eq,\beta_{2f}^{(i)}(t)} \end{bmatrix}, \quad A_{eq,\xi_2^{(i)}(t)} = \begin{bmatrix} 1 & \cdots & 1 & 0 & \cdots & 0 \end{bmatrix}, \tag{4.57}$$

$$A_{eq,\beta_{2b}^{(i)}(t)} = \begin{bmatrix} 0 & \cdots & 0 & 1 & \cdots & 1 & 0 & \cdots & 0 \\ \vdots & \ddots & \vdots & \vdots & \ddots & \vdots & \ddots & \vdots \\ 0 & \cdots & 0 & 0 & \cdots & 0 & 1 & \cdots & 1 \end{bmatrix}, \tag{4.58}$$

$$A_{eq,\beta_{2f}^{(i)}(t)} = \begin{bmatrix} -1 & \cdots & 0 & 1 & \cdots & 0 & \cdots & 1 & \cdots & 0 \\ \vdots & \ddots & \vdots & \vdots & \ddots & \vdots & \ddots & \vdots & \ddots & \vdots \\ 0 & \cdots & -1 & 0 & \cdots & 1 & \cdots & 0 & \cdots & 1 \end{bmatrix}, \tag{4.59}$$

The vector b_{eq} has the following form:

$$b_{eq} = \begin{bmatrix} b_{eq,a} & b_{eq,r} \end{bmatrix}^T, \tag{4.60}$$

$$b_{eq,c} = \begin{bmatrix} b_{eq,\Theta(1)} \\ \vdots \\ b_{eq,\Theta(T)} \end{bmatrix}, \quad b_{eq,\Theta(t)} = \begin{bmatrix} b_{eq,\Psi^{(1)}(t)} \\ \vdots \\ b_{eq,\Psi^{(N)}(t)} \end{bmatrix}, \quad b_{eq,\Psi^{(i)}(t)} = \begin{bmatrix} b_{eq,\xi^{(i)}(t)} \\ b_{eq,\beta^{(i)}(t)} \end{bmatrix}, \tag{4.61}$$

$$b_{eq,\xi^{(i)}(t)} = \begin{bmatrix} 1 \end{bmatrix}, \quad b_{eq,\beta^{(i)}(t)} = \begin{bmatrix} 0 \\ \vdots \\ 0 \end{bmatrix}, \tag{4.62}$$

$$lb = \begin{bmatrix} 0 \\ \vdots \\ 0 \end{bmatrix}, \quad ub = \begin{bmatrix} 1 \\ \vdots \\ 1 \end{bmatrix} \tag{4.63}$$

The matrix A_{ineq} is given by:

$$A_{ineq} = \begin{bmatrix} A_{ineq,\Theta(1)} \\ \vdots \\ A_{ineq,\Theta(T)} \end{bmatrix}, \quad A_{ineq,\Theta(t)} = \begin{bmatrix} A_{ineq,\Psi^{(1)}(t)} \\ \vdots \\ A_{ineq,\Psi^{(N)}(t)} \end{bmatrix}, \tag{4.64}$$

$$A_{ineq,\Psi^{(i)}(t)} = \begin{bmatrix} A_{ineq,\Psi^{(i)}(t),a} & A_{ineq,\Psi^{(i)}(t),r} \end{bmatrix}, \quad A_{ineq,\Psi^{(i)}(t),r} = -A_{ineq,\Psi^{(i)}(t),a} \tag{4.65}$$

$$A_{ineq,\Psi^{(i)}(t),a} = \begin{bmatrix} 0 & \cdots & 0 & 1 & \cdots & 0 & 0 & \cdots & 0 \\ 0 & \cdots & 0 & -1 & \cdots & 0 & 0 & \cdots & 0 \\ \vdots & \ddots & \vdots & \vdots & \ddots & \vdots \\ 0 & \cdots & 0 & 0 & \cdots & 1 & 0 & \cdots & 0 \\ 0 & \cdots & 0 & 0 & \cdots & -1 & 0 & \cdots & 0 \end{bmatrix} \tag{4.66}$$

Finally, the vector b_{ineq} is given by:

$$b_{ineq} = \begin{bmatrix} b_{ineq,\Theta(1)} \\ \vdots \\ b_{ineq,\Theta(T)} \end{bmatrix}, \quad b_{ineq,\Theta(t)} = \begin{bmatrix} b_{ineq,\Psi^{(1)}(t)} \\ \vdots \\ b_{ineq,\Psi^{(N)}(t)} \end{bmatrix}, \quad b_{ineq,\Psi^{(i)}(t)} = \begin{bmatrix} \alpha \\ \vdots \\ \alpha \end{bmatrix} \tag{4.67}$$

As described in Sect. 4.3, the dimensionality of the variables vector and, accordingly, of each elements of the QP problem is defined as follows:

- v_c: l-dimensional vector;
- H_c: $[l \times l]$ symmetric matrix;
- f_c: l-dimensional vector;
- $A_{eq,c}$: $[m \times l]$ matrix;
- $b_{eq,c}$: m-dimensional vector;
- lb, ub: $2l$-dimensional vector;
- A_{ineq}: $[2l \times 2l]$ matrix;
- b_{ineq}: $2l$-dimensional vector;

where $l = T \cdot \sum_{i=1}^{N}(m_i + m_i^2)$ and $m = T \cdot \sum_{i=1}^{N}(1 + 2m_i)$.

4.3.2 Experimental Setup

The proposed approach has been compared with the algorithm presented by Hart in [15], since it employs both the active and the reactive power to model the appliance working behaviour and it employs those electrical parameters for disaggregation. This section provides an overview of its basic operating principles as well as additional details on its implementation. In addition, the algorithm originally presented in [15] has been improved for handling the occurrence of multiple solutions by means of a MAP technique.

Hart's algorithm models each appliance as a Finite State Machine (FSM). Each FSM is represented by the following parameters:

- the number of states $m \in \mathbb{Z}_+$;
- the finite states $x \in \{S_1, S_2, \ldots, S_m\}$;
- the symbols emitted $\boldsymbol{\mu}_j \in \mathbb{R}^n$, where $j = 1, \ldots, m$;
- state transition matrix $\boldsymbol{T} \in \{0, 1\}^{m \times m}$.

As in the proposed approach, each state of the FSM corresponds to a working state of the appliance and $n = 2$, i.e., the symbol emitted in the j-th state is defined as $\boldsymbol{\mu}_j = [\mu_{a,j} \, \mu_{r,j}]^T$. A tolerance parameter $\boldsymbol{\beta}_j = [\beta_{a,j} \, \beta_{r,j}]^T$ is associated to the emitted symbol in the j-th state, in order to define the effectiveness interval for the emitted symbol. The interval width is $2\boldsymbol{\beta}_j$ and it is centred in $\boldsymbol{\mu}_j$. For each appliance, the quantities to be estimated are the number of states m, the values of $\boldsymbol{\mu}_j$ and $\boldsymbol{\beta}_j$ for each state and the state transition matrix $\boldsymbol{T}$.

In order to model the power consumption of an appliance as a stochastic process, under the assumption of multiple independent causes to the circuital power dissipation, the central limit theorem might be invoked. Therefore, the power consumption $y^{(i)}(t)$ of the i-th appliance at time instant t, related to the working state $x^{(i)}(t)$, can be modelled as a bivariate Gaussian variable, described by a mean vector $\boldsymbol{\mu}_{x^{(i)}(t)}$ and a covariance matrix $\boldsymbol{\Sigma}_{x^{(i)}(t)}$:

$$y^{(i)}(t)|x^{(i)}(t) \sim \mathcal{N}\left(\boldsymbol{\mu}_{x^{(i)}(t)}, \boldsymbol{\Sigma}_{x^{(i)}(t)}\right). \tag{4.68}$$

Following this approach, the consumption signal is replaced by a simplified model that represents a constant power consumption, corresponding to the mean value of the working state power value, with a superimposed noisy contribution, described by the variance value in the working state. Under the assumption of statistical independence between the active and reactive power components, the covariance matrix $\boldsymbol{\Sigma}_{x^{(i)}(t)}$ is diagonal:

$$\boldsymbol{\Sigma}_{x^{(i)}(t)} = \begin{bmatrix} \sigma^2_{a,x^{(i)}(t)} & 0 \\ 0 & \sigma^2_{r,x^{(i)}(t)} \end{bmatrix}, \tag{4.69}$$

where $\sigma^2_{a,x^{(i)}(t)}$ and $\sigma^2_{r,x^{(i)}(t)}$ represent, respectively, the variance of the active and reactive power in the cluster. The inference procedure is carried out independently for the two components. Therefore, at each state,

$$y^{(i)}_c(t)|x^{(i)}(t) \sim \mathcal{N}\left(\mu_{c,x^{(i)}(t)}, \sigma^2_{c,x^{(i)}(t)}\right). \tag{4.70}$$

The number of states m_i is defined in the clustering phase, described in Sect. 4.1.1, assuming that each cluster corresponds to a state in the FSM model: the estimation of the mean and the variance values for each component is performed with the Maximum Likelihood criterion on the clusters data. Each component of the

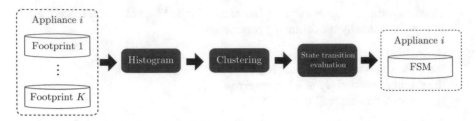

Fig. 4.14 Block diagram of the clustering and of the model training stages of Hart's algorithm

tolerance parameter $\beta_{c,j}$, associated to the respective component of the emitted symbol $\mu_{c,j}$, is set equal to the standard deviation $\sigma_{c,j}$ of the Gaussian distribution.

Regarding the state transition matrix T, each entry T_{ij} represents the admissibility of the transition from state i to state j, using the value $T_{ij} = 1$ if the transition is allowed and $T_{ij} = 0$ otherwise. This value is inferred from the ground truth state evolution of each appliance consumption. Since this model does not represent the evolution in time of a signal, the permanence in the state is not represented, therefore the variable T_{ii} is set to 1. The diagram of the clustering and of the model training stage is shown in Fig. 4.14.

Since the aggregated data $\overline{y}_c(t)$ is assumed to correspond with the sum of the power consumption of each appliance, it can be modelled as a Gaussian variable, described by a mean value and a variance value equivalent to the sum of the corresponding values of each appliance, under the assumption of statistical independence among the appliances:

$$\overline{y}_c(t)|x^{(1:N)}(t) \sim \mathcal{N}\left(\sum_{i=1}^{N} \mu_{c,x^{(i)}(t)}, \sum_{i=1}^{N} \sigma^2_{c,x^{(i)}(t)}\right). \tag{4.71}$$

This variable represents the Probability Density Function (PDF) of the working states combinations and it allows to evaluate which combination of working states fits the power value for each sample of the aggregated data. The number of admissible combinations of working states is equal to $\prod_{i=1}^{N} m_i$.

Following the same rule defined for each appliance symbol, the effectiveness interval for each combination is centred in mean value, and its width is twice the value of the standard deviation. For some combinations, which have similar mean value or great variance, the effectiveness intervals are overlapped: for those cases, if the power value falls in this region, both the combinations are considered valid.

The aggregate power data is analysed sample by sample: for each value, the effectiveness intervals in which the sample falls are selected. The related state combination might be admissible or not, depending on the previous state combination selected. Therefore, for each FSM, from the knowledge of the previous state selected, the admissible transition is evaluated through the transition matrix T_{ij}: the FSMs which do not make any variation in the state from the previous combination are not evaluated, then if the transition is not admissible for at least one

FSM, the selected combination is discarded. The starting combination is evaluated on the first sample, without the evaluation on the transition from any previous state. If no combination is admissible, the previous state is maintained for each FSM. If the aggregated data sample does not fall within any combination interval, the previous state is maintained for each FSM.

In this way, the time series of the state evolution is reconstructed for each FSM. The disaggregation consists in using the related power level consumption assigned to each state of the FSM, thus reconstructing the power consumption profile for each appliance. The general scheme of the disaggregation phase is shown in Fig. 4.15.

In order to deal with the noise presence in the aggregated data, an FSM version of the *noise* model defined in Sect. 4.1.2 is considered, additionally to the FSM models representing the appliances.

In order to make a fair comparison of algorithms, representing an appliance, both kinds of model have the same number of states, values of power consumption and standard deviation of the gaussian variable. The values are resumed in Table 4.3.

In [15], the author did not describe the technique adopted for dealing with the occurrence of multiple solutions during the disaggregation phase. Two different approaches for dealing with the problem are adopted. The first consists in supposing

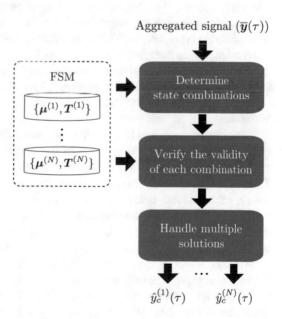

Fig. 4.15 Diagram of the load disaggregation phase

Table 4.3 Number of states m_i related to each class of appliance

Problem dimensionality	Dryer	Washing machine	Dishwasher	Fridge	Electric oven	Heat pump
Univariate	3	4	3	2	3	3
Bivariate	3	5	4	2	4	3

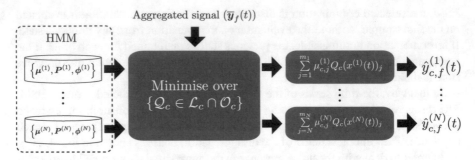

Fig. 4.16 Diagram of the load disaggregation phase

that each combination of appliances is equally probable, thus the ambiguity is solved by choosing a random combination sampled from a uniform distribution. This algorithm will be denoted as "Hart" in the remainder of this book.

The second approach consists in adopting a MAP technique [58]: the *posterior probability* of each combination is calculated from the training data, and it is multiplied to the Gaussian PDF, resulting in the *posterior PDF*. The value of the posterior PDF in the aggregate data sample is denoted as the *posterior likelihood*. The combination with the higher posterior likelihood value is then chosen as the most probable combination. This alternative of Hart's algorithm will be denoted as "Hart w/ MAP" in the remainder of this book.

The general scheme of the disaggregation phase is shown in Fig. 4.16. The algorithm is based on the work proposed by Kolter and Jaakkola [21], where the problem is modelled in the Additive Factorial Hidden Markov Model (AFHMM) framework.

Basically, this consists in modelling the value of each aggregated power sample as a combination of working states of the appliances. In [21], an assumption is made that at most one HMM changes its state at any given time, which holds true if the sampling time is reasonably short. In this case, a transition on the aggregate power, when moving from a sample to the next, corresponds to a state change of a particular HMM. As a consequence, a differential signal can be modelled as the result of a Differential Factorial Hidden Markov Model (DFHMM), which relies on the same HMM models comprising the AFHMM. The DFHMM models the observation output as the difference between the states combination of the HMMs in two consecutive time instants. By combining the additive and differential models, the inference on the set of states of multiple HMMs can be computed through the Maximum A Posteriori (MAP) algorithm, which takes the form of a Mixed Integer Quadratic Programming (MIQP) optimization problem. One of the shortcomings of this approach is the non-convex nature of the problem, due to the integer nature of the variables: therefore, a relaxation towards real values is taken into account, which allows the solution to assume any value in the range [0, 1], instead of the binary solution, leading to a convex Quadratic Programming (QP) optimization problem.

In a real case scenario, the modelled output may not match with the observed aggregated signal, due to electrical noises, very small loads or leakages. In that case, the issue is addressed by defining a robust mixture component both in the AFHMM and in the DFHMM. This component is missing in this book, since all the contributions to the aggregated power are modelled. Indeed, each appliance and the *noise* is represented by its HMM.

The dataset used for the experiments is the Almanac of Minutely Power dataset (AMPds) [58]: it contains recordings of consumption profiles belonging to a single home in Canada for a period of 2 years, at 1 min sampling rate. Additionally to the aggregated power consumption, it provides active and reactive power at appliance level, unlike most of the dataset, in which the appliances consumption is described by the only active power, as showed in Sect. 2.3: this information is crucial in order to create the appliance models and test the new approach.

The experiments are conducted by using the six appliances which contribute the most to the power consumption: dryer, washing machine, dishwasher, fridge, electric oven and heat pump. Regarding the significance of the reactive components of the appliances taken into consideration, the following values have been extracted from the datasets: (128.25 W, 7.96 VAR) for the fridge, (4545.91 W, 413.75 VAR) and (248.11 W, 408.94 VAR) for the dryer, (909.11 W, 203.44 VAR), (531.10 W, 14.37 VAR),(146.80 W, 3.60 VAR) and (137.54 W, 96.47 VAR) for the washing machine, (753.07 W, 33.31 VAR), (137.96 W, 35.86 VAR) and (14.42 W, 52.55 VAR) for the dishwasher, (3187.67 W, 136.63 VAR),(125.68 W, 121.67 VAR) and (89.54 W, 50.62 VAR) for the electric oven, (1798.83 W, 320.95 VAR) and (37.23 W, 17.03 VAR) for the heat pump. As shown by these values, the appliances evaluated in the experiments have a significant contribution of reactive power that make them suitable for evaluating the performance of the proposed approach. Analysing the contents of the dataset, the usage of the appliances proves to be homogeneous throughout the entire period, therefore the experiments are evaluated on 6 months of data, which can be considered a representative of the entire dataset. A subset of the data, spanning over 14 days, has been considered sufficient to collect all the signatures required to train all the HMMs. This represents the training set in Fig. 4.5b.

Two different scenario are defined in this work, according to [87]. The *noised* scenario employs the aggregated power consumption in the dataset as the aggregated signal, therefore it includes the noise term. In this case, the training data used to create the *noise model* are obtained subtracting the ground truth consumption signals, related to the appliances of interest, from the aggregated power, whereas in the *denoised* scenario the aggregated data are synthetically composed by summing the ground truth appliance power signals in the dataset, determining the absence of the noise term.

The proposed approach and Hart's algorithm are able to disaggregate both the active and the reactive power, however the performance metrics has been calculated on the active power only in order to compare it with the univariate formulation of AFAMAP. Furthermore, the active power is the physical quantity directly related to the cost in the bill, therefore it is the most relevant component to be analysed.

The frame size is set to $T = 60 \, min$, which is an interval sufficiently large to include a complete activation for the most of appliances under study. This value is considered within the *windowing* operation in Fig. 4.16, where the f-th frame is considered in the disaggregation. For the ones which have a longer activation, this value allows to include a complete operating subcycle, for which the HMM is still representative. The variance parameters are set to $\sigma_{c,1}^2 = \sigma_{c,2}^2 = 0.01$ according to the variance of the experimental data, and the regularization parameter is set to $\lambda = 1$.

The algorithm has been implemented in Matlab and the CPLEX[1] solver has been used to solve the QP problem. The amount of time required to disaggregate a frame of 60 min on a personal computer equipped with an Intel i7 CPU running at 3.3 GHz and 32 GB of RAM is about 30 s. The performance is compared to the univariate formulation of AFAMAP and to Hart's algorithm presented in Sect. 4.3.2. The tolerance parameter is set $\alpha = 10^{-6}$.

Table 4.3 presents the number of states, defined a-priori for each class of appliance. For appliances with similar consumption value in active power, different values of reactive power are associated: this phenomenon allows to reduce the number of state combination in the aggregate power, when passing from the univariate to the bivariate approach, improving the disaggregation performance.

The number of states in the *noise* model has been varied in the range $\{4, 6, 8, 10\}$, both in the univariate and bivariate approaches, in order to find the most performing model.

4.3.3 Results

In this section, the results of the experiments related to the *denoised* scenario will be shown. Since the aggregated power signal depends on which and how many appliances are considered, the experiments have been conducted by varying the number of appliances, in order to evaluate the disaggregation performance for different problem complexities. In particular, different test sets, each composed of every combination of N appliances have been created. For each test set, the total number of experiments is $\binom{6}{N}$, with $N = 2, \ldots, 5$ and the final metrics are calculated averaging between the single experiments overall performance. Before calculating the final energy based F_1-Measure ($F_1^{(E)}$), the Precision ($P^{(E)}$) and Recall ($R^{(E)}$) are averaged between the experiments. Differently, the final NDE is the average between the single experiment value.

In Fig. 4.17, the disaggregated appliances active power (D) are compared to the corresponding ground truth (GT): in the figure, for each appliance, an adequate time span is considered, in order to evaluate the performance on a single or multiple activations. The bottom of the figure shows the comparison of the appliance

---

[1]http://www-01.ibm.com/software/commerce/optimization/cplex-optimizer/.

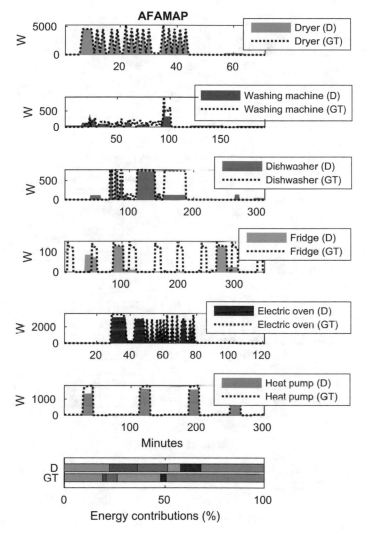

Fig. 4.17 Algorithms comparison: AFAMAP vs Hart vs proposed approach. For each algorithm, the disaggregation output (D) is compared against the ground truth (GT) signals

contribution to the total energy in the aggregated signal, between the disaggregation outputs and the ground truth consumptions. The left side of the figure shows the disaggregation profiles resulting from the univariate formulation of the AFAMAP algorithm, the central shows the active power component resulting from the Hart's algorithm, and the right side shows profiles related to the proposed approach (Table 4.4).

The overall disaggregation results are reported in Fig. 4.18, where the $F_1^{(E)}$ is reported in Fig. 4.18a and the NDE in Fig. 4.18b. The values are related to

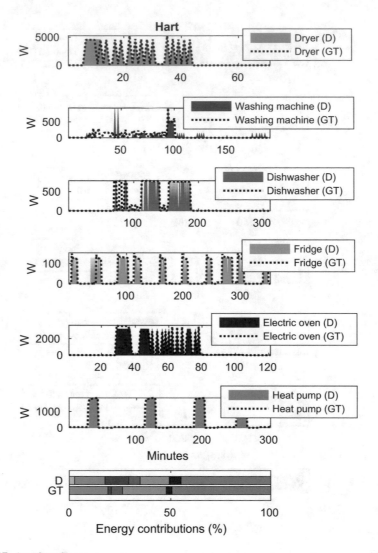

Fig. 4.17 (continued)

Table 4.5, where the absolute improvements of the proposed approach with respect to the AFAMAP and the Hart's algorithm are shown. The proposed approach reaches the best performance in each case study, with $F_1^{(E)}$ of 87.0 and NDE equal to 0.209 in the 2 appliances case, and with $F_1^{(E)}$ of 69.4 and NDE equal to 0.347 in the 6 appliances case, The proposed approach reaches the best performance in each case study, with $F_1^{(E)}$ of 87.0 and NDE equal to 0.209 in the 2 appliances case, and with $F_1^{(E)}$ of 69.4 and NDE equal to 0.347 in the 6 appliances case.

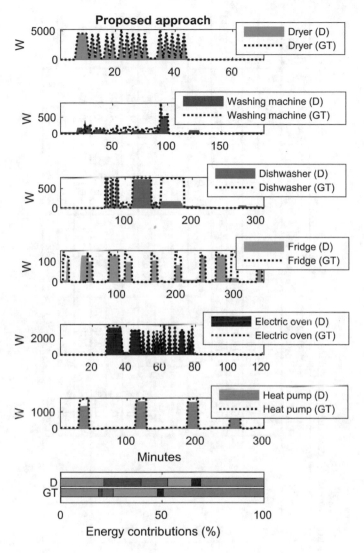

Fig. 4.17 (continued)

The radar chart in Fig. 4.19 shows the $F_1^{(E)}$ for each appliance. It refers to the experiment including all the 6 appliances, where the area of each coloured line is proportional to the $F_1^{(E)}$ of the related algorithm, averaged across the appliances. The values are related to Table 4.4, where the absolute improvements of the proposed approach with respect to the AFAMAP and the Hart's algorithm are shown.

As shown in the plots, the appliances presenting a high steady power consumption are easily recognized, whereas the appliances with complex working cycles, or

Table 4.4 Performance improvement in the "6 appliances" case study (denoised scenario)

Algorithm	Metric	Dryer	Washing machine	Dishwasher	Fridge	Electric oven	Heat pump
AFAMAP	(I) $F_1^{(E)}$ (%)	87.3	14.5	44.4	35.5	38.0	76.9
Hart	(II)	54.9	10.4	33.1	76.4	32.1	73.4
Hart w/ MAP	(III)	86.8	14.0	57.5	74.4	54.5	90.8
Proposed approach	(IV)	90.2	13.8	57.1	58.9	71.5	76.5
Improvement	(IV)–(I) $\Delta F_1^{(E)}/F_1^{(E)}$ (%)	+3.3	−4.8	+28.6	+65.9	+88.2	−0.5
	(IV)–(II)	+64.3	+32.7	+72.5	−22.9	+122.7	+4.2
	(IV)–(III)	+3.9	−1.4	−0.7	−20.8	+31.2	−15.8
AFAMAP	(I) NDE	0.215	2.279	0.685	0.878	0.478	0.388
Hart	(II)	0.798	4.383	1.282	0.670	1.725	0.739
Hart w/ MAP	(III)	0.481	3.003	0.768	0.685	1.026	0.411
Proposed approach	(IV)	0.229	2.384	0.446	0.735	0.286	0.377
Improvement	(IV)–(I) ΔNDE	+0.014	+0.104	−0.239	−0.143	−0.192	−0.011
	(IV)–(II)	−0.569	−1.999	−0.836	+0.065	−1.439	−0.363
	(IV)–(III)	−0.252	−0.620	−0.322	+0.049	−0.740	−0.034

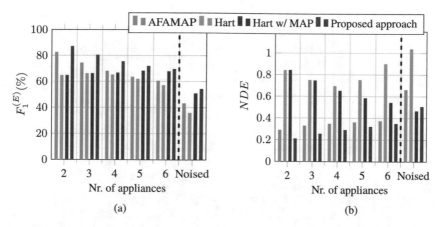

Fig. 4.18 Disaggregation performance on AMPds dataset for all the addressed algorithms. (a) Comparison of the disaggregation performance in terms of $F_1^{(E)}$ for different number of appliances. (b) Comparison of the disaggregation performance in terms of NDE for different number of appliances

with several power levels, are more difficult to detect. For instance, the dryer, the electric oven and the heat pump are successfully reconstructed, whereas the washing machine, the dishwasher and the fridge are partially erroneously reconstructed. Indeed, in the univariate formulation, whenever several appliances present similar consumption levels, many combinations may satisfy the problem constraints and the algorithm chooses an erroneous solution for disaggregation. Comparing the results with the proposed bivariate approach, the multiple combinations of the solution are reduced due to the component constraint to be satisfied by the algorithm, which leads to the correct solution and, consequently, to a better profile disaggregation of the active power component. For instance, although the appliances with higher power level maintain a successful disaggregation, the fridge and the dishwasher improve the correspondence with the ground truth signals. The washing machine partially improves the disaggregation performance in the activation period, whereas introduces some false energy assignation. The disaggregated profiles of Hart's method show that, for some appliances, the FSM is a modelling technique which allows a better representation for the appliances with sharply defined steady states, e.g., the fridge and the heat pump, but a worse representation for appliances with highly variable activity, e.g., the electric oven.

The more confident are the disaggregated profiles with respect to the ground truth signal, the better is the estimation of the energy consumption percentage distribution among the appliances: indeed, for the proposed approach, the consumption distribution has a better correspondence with the ground truth ones, with respect to the AFAMAP algorithm. For instance, the disaggregated profiles related to the fridge results to be more confident, which reflects on the increase of the energy assignation, whereas the dishwasher and the electric oven ones results to have a false energy assignation during the OFF period, corresponding to a decrease of

Table 4.5 Comparison of the disaggregation performance for different number of appliances (denoised scenario)

Algorithm	Metric	2 appl.	3 appl.	4 appl.	5 appl.	6 appl.
AFAMAP (I)	$F_1^{(E)}$ (%)	82.4	74.2	68.0	63.5	60.4
Hart (II)		64.5	66.0	65.1	61.8	57.0
Hart w/ MAP (III)		64.6	66.1	66.6	68.1	67.7
Proposed approach (IV)		**87.0**	**80.3**	**75.3**	**71.8**	**69.4**
Improvement (IV)–(I)	$\Delta F_1^{(E)}/F_1^{(E)}$ (%)	+5.6	+8.2	+10.7	+13.1	+14.9
(IV)–(II)		+34.9	+21.7	+15.7	+16.2	+21.8
(IV)–(III)		+34.7	+21.5	+13.1	+5.4	+2.5
AFAMAP (I)	NDE	0.288	0.327	0.346	0.360	0.371
Hart (II)		0.839	0.748	0.693	0.750	0.899
Hart w/ MAP (II)		0.840	0.744	0.650	0.582	0.541
Proposed approach (IV)		**0.209**	**0.254**	**0.289**	**0.319**	**0.347**
Improvement (IV)–(I)	ΔNDE	−0.079	−0.073	−0.057	−0.041	−0.024
(IV)–(II)		−0.630	−0.494	−0.404	−0.431	−0.552
(IV)–(III)		−0.631	−0.490	−0.361	−0.263	−0.194

Bold values represent the higher performance in the algorithm comparison. Thus, the algorithm with bold values is the best algorithm in the experiments

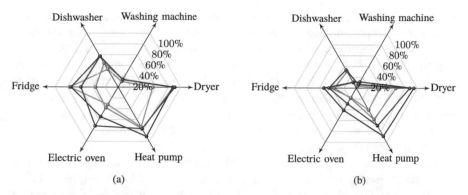

Fig. 4.19 Performance in terms of $F_1^{(E)}$ (%) for the different appliances in the "6 appliances" case study: (**a**) denoised scenario, (**b**) noised scenario

the related energy contributions. Regarding the washing machine, some errors are introduced, therefore the energy assignation is erroneously increased. Regarding the dryer and the heat pump the energy contributions are maintained, because of the correspondence between the algorithms disaggregation performance. In the Hart's method, the improvements in the heat pump and the fridge are reflected on a better correspondence between the energy contributions, but the absence of the constraint between the aggregate power amount and the sum of the disaggregated profiles leads to an unassigned percentage of the total energy (represented as the *grey* portion).

Regarding the performance of the individual appliances, the major improvements with respect to AFAMAP are observed in the electric oven, the fridge and the dishwasher, with a relative increase of the $F_1^{(E)}$ of +88.2%, +65.9% and +28.6%, and a variation in the NDE of −0.192, −0.143, −0.239, respectively. This is due to a more accurate correspondence between the disaggregated output and the ground truth, as already shown in the disaggregation output plots. On the contrary, the performance is almost unchanged for the washing machine, the dryer and the heat pump. With respect to the Hart's algorithm, the proposed approach shows a high improvement additionally for the dryer, with an absolute increase of $F_1^{(E)}$ equal +64.3% and a variation in the NDE of −0.569, whereas it shows a substantial loss for the fridge, with a decrease of $F_1^{(E)}$ equal −22.9% and a variation in the NDE of +0.065. This demonstrates that the HMM modelling results more effective with a higher number of states. Since moving from the univariate to the bivariate model leads to a greater number of states, this also demonstrates the effectiveness of the proposed approach. Compared to the Hart's algorithm with the MAP stage, the performance on each appliance reduces their gain, particularly for the dishwasher and the dryer, with a decrease of $F_1^{(E)}$ equal to −0.7% and an increase of +3.9% and a variation in the NDE of −0.322 and −0.252 up to the heat pump, where a loss of performance is shown, with an absolute increase in the $F_1^{(E)}$ of −15.8% and a variation in the NDE of −0.034. The washing machine remains the appliance with the worst disaggregation performance: the reason is the model complexity, since it is the appliance with the highest number of states, both in the univariate and bivariate representation. Observing the radar chart, the area under the curve related to the proposed approach is increased with respect to AFAMAP and Hart's algorithm, resulting in an average performance improvement, whereas it is slightly higher with respect to the Hart's algorithm version with the MAP stage. The average performance of the system increases, resulting in a relative improvement of $F_1^{(E)}$ equal to $+14.9\%$, $+21.8\%$ and $+2.5\%$, and a variation in the NDE of -0.024, -0.552, -0.194 with respect to AFAMAP, the Hart's algorithm and the version with MAP stage, respectively.

Concerning with the experiments for different number of appliances, the results show that, lowering the number of appliances, the performance improves in the FHMM-based algorithms, while in the Hart's algorithm it reaches a peak with 4 appliances, after that the performance decreases. Regarding the Hart's algorithm version with the MAP stage, the performance decreases gradually with a lower number of appliance.

Compared to AFAMAP and to Hart's algorithm, the proposed approach provides a significant performance improvement also when the problem complexity is minimal, i.e., when the number of appliances is 2. The higher absolute increase from AFAMAP occurs with 6 appliances, whereas it decreases lowering the complexity of the problem: this demonstrates that the proposed approach resolves more ambiguities in the NILM solution when the number of combinations of working states is higher.

Regardless of the number of appliances, the performance of Hart's algorithm is lower compared to the proposed approach, because of the less descriptive capabilities of the FSM appliance model with respect to the HMM one. The comparative evaluation with the Hart's version with the MAP stage proves that, even if this approach exploits the information on the most probable solution in case of ambiguity, which is an ideal condition, the proposed approach reaches better performance. Furthermore, the proposed algorithm provides an optimum solution on a frame of T samples, which takes into account both the short-term and long-term dependencies of the signal. This differs in Hart's algorithm that finds the solution by processing the aggregate signal sample-by-sample. For this method, the performance decreases reducing the number of the appliances: a motivation behind this phenomenon can reside in the fact that the MAP stage of the Hart's algorithm chooses a solution with higher probability, but which results incorrect for the majority of the experiments, specially with few combinations.

In this section, the results of the experiments related to the *noised* scenario will be shown. Differently from the *denoised* scenario, the aggregated power signal does not vary with the appliances considered, therefore only the results with all the appliances will be shown. Regarding the number of states of the *noise model*, the experiments demonstrated that, for each approach, the best value is 4, except for the Hart's algorithm with the MAP stage, for which the best results are reached with 10 states. For the sake of conciseness, only the results for the best configuration will be reported in this section.

The overall disaggregation results are reported in Fig. 4.18, on the last column, in order to make a comparative evaluation with the *denoised* scenario. The values are related to Table 4.6 on the *Overall* column, where the absolute improvements of the proposed approach with respect to the AFAMAP and the Hart's algorithm are shown. The proposed approach reaches the best overall performances, with $F_1^{(E)}$ of 54.1 and NDE equal to 0.504, despite the Hart's algorithm version with the MAP stage showing a higher NDE value. This discordance will be motivated in the analysis. The radar chart in Fig. 4.19b shows the $F_1^{(E)}$ for each appliance. The values are related to Table 4.6.

Differently from the *denoised* scenario, the major improvement, with respect to AFAMAP, is observed for the dryer, with an $F_1^{(E)}$ relative improvement of $+35.9\%$, and a variation in the NDE of -0.033, whereas the improvements are reduced for the remaining appliances. This proves the effectiveness of the transition from the univariate to the bivariate formulation of the problem, even in the presence of noise.

With respect to Hart's algorithm, the proposed approach shows a higher improvement for the dryer, the dishwasher and the heat pump with an improvement of $+131.7\%$, $+228.2\%$, $+71.7\%$, and a variation in the NDE of -0.610, -0.884, -0.462. Differently, Hart's algorithm with the MAP stage achieves a higher $F_1^{(E)}$, and the relative difference of $F_1^{(E)}$ for the heat pump, the electric oven and the dryer is -19.4%, -12.2%, -6.8%, while in terms of NDE the difference is $+0.036$, -0.044, $+0.019$. This demonstrates that the HMM modelling leads to performance improvements with respect to the FSM modelling even in the presence of noise, but considering the MAP stage this improvement is substantially reduced. The washing machine is still the appliance with the worst disaggregation performance, following the trend of the *denoised* scenario. Observing the radar chart, the area under the curve related to the proposed approach is increased with respect to AFAMAP and Hart's algorithm, resulting in an average performance improvement, whereas it is comparable with respect to the Hart's algorithm version with the MAP stage, due to unbalancing between the appliances.

The average performance of the system increases, resulting in an $F_1^{(E)}$ absolute improvement of $+25.5\%$, $+51.1\%$ and $+6.7\%$, and a variation in the NDE of -0.155, -0.533, $+0.040$ with respect to AFAMAP, the Hart's algorithm and the version with MAP stage, respectively.

Comparing those results to the *denoised* scenario ones, the overall performance is lower, due to the introduction of the noise contribution in the aggregated power, except for the Hart's algorithm with the MAP stage: despite the $F_1^{(E)}$ showing a degradation of performance, the NDE decreases, meaning that this version of the algorithm maintains the trend showed with the increase of the number of appliances. In fact, the *noised* scenario can be defined as the *denoised* scenario using the *noise* model additionally to the appliances models, therefore the MAP stage introduces additional advantages, leading to a performance improvement. The MAP stage exploits additional information which are not introduced within the AFHMM, but represents an almost ideal FSM based case study.

Table 4.6 Appliances performance improvement in the "6 appliances" case study (noised scenario)

Algorithm	Metric	Dryer	Washing machine	Dishwasher	Fridge	Electric oven	Heat pump	Overall
AFAMAP (I)	$F_1^{(E)}$ (%)	64.6	6.3	30.7	35.6	18.7	57.7	43.1
Hart (II)		37.9	3.7	10.3	42.1	17.6	40.3	35.8
Hart w/ MAP (III)		94.6	11.6	10.7	52.7	41.1	88.6	50.7
Proposed approach (IV)		87.8	6.9	33.8	44.6	28.9	69.2	**54.1**
Improvement (IV)–(I)	$\Delta F_1^{(E)}/F_1^{(E)}$ (%)	+35.9	+9.5	+10.1	+25.3	+54.5	+19.9	+25.5
(IV)–(II)		+131.7	+86.5	+228.2	+5.9	+64.2	+71.7	+51.1
(IV)–(III)		−7.2	−40.5	+215.9	−15.4	−29.7	−21.9	+6.7
AFAMAP (I)	NDE	0.305	4.395	0.888	0.909	0.939	0.787	0.659
Hart (II)		0.882	5.960	1.714	0.982	1.593	0.929	1.037
Hart w/ MAP (III)		0.254	1.965	1.110	0.942	0.974	0.432	**0.464**
Proposed approach (IV)		0.272	4.055	0.829	0.861	0.930	0.467	0.504
Improvement (IV)–(I)	ΔNDE	−0.033	−0.340	−0.058	−0.048	−0.010	−0.319	−0.155
(IV)–(II)		−0.610	−1.905	−0.884	−0.121	−0.663	−0.462	−0.533
(IV)–(III)		+0.019	+2.090	−0.280	−0.081	−0.044	+0.036	+0.040

Bold values represent the higher performance in the algorithm comparison. Thus, the algorithm with bold values is the best algorithm in the experiments

4.4   Footprint Extraction Procedure

Among different NILM approaches, the supervised ones reach better performance [52, 55], that is, the resulting disaggregated signals have a better correspondence with the true appliance energy consumption. Therefore, those methods results to be more reliable for the final user.

The supervised section in the NILM algorithms corresponds to the appliance modelling stage, as showed in Fig. 4.20b, where the training phase is carried out. A model is created starting from the appliance level consumption (e.g., training set), in order to represent each appliance in a parametric way, and its parameters are used in the NILM algorithm in order to disaggregate the portion of the aggregated power consumption related to each appliance, as represented in Fig. 4.20c.

Fig. 4.20 The supervised NILM chain. (**a**) The footprint extraction stage. (**b**) The appliance modelling stage. (**c**) The disaggregation algorithm stage

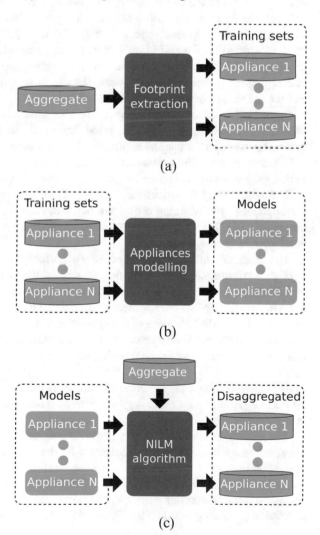

The power consumption profile of an appliance can be depicted as the repeating of a working cycle, alternated by time intervals when the appliance is turned off. The repetition rate, related to the length of the *off-intervals*, depends on the user consumption habit.

Therefore, in order to analyze the consumption features of an appliance, it is sufficient to extract the working cycle in the appliance level consumption, defined as the *footprint*, and to exploit it as training set in the appliance modelling stage.

This stage of the supervised NILM chain is named *footprint extraction*, as showed in Fig. 4.20a.

In literature, different approaches have been proposed to extract the appliance working cycle features from the aggregated data. An unsupervised method, based on spectral clustering, is proposed in [21]: the most different activation occurrences, which can be denoted in the aggregated power, are saved; then, they are grouped between the most similar, using the clustering technique. A bayesian approach is used in [18, 19]: a generic bayesian model for the appliance category is defined; then, it is fitted on the activation within the aggregated power, using a threshold schema on the likelihood function. Most of those approaches have limitations, concerning the aggregated power, where the appliance activation can be overlapped and it can cause trouble in the extraction phase.

To overcome this, in a real scenario, the user interaction with the system can be considered, in order to improve the reliability of the footprint extraction: in those cases, the user needs a facilitated procedure to determinate the appliance activation instant and an easy way to interact with the energy monitoring system. Therefore, in this work a *user-aided* footprint extraction procedure is proposed.

The easiest way to extract the footprint from the aggregated power is to use the appliance alone, turning off all the other devices in the electrical network, as described in [15]. This approach results to be the more reliable for the user, thus it is adopted in the presented work.

The appliance modelling stage employs the footprint, in order to represent the appliance consumption behaviour: despite several works dealing with model for the classification, such as SVM, k-NN [36] or deep neural networks [31], the hidden Markov Model (HMM) is a widespread modelling technique [17, 22, 28], since it is able to represent the behaviour of the appliance in working states and to regulate the transition with a probability value. This representation is close to the real appliance mode of operation, where each working state corresponds to a power consumption value.

In this work, the disaggregation algorithm is based on HMM, in particular the AFAMAP (Additive Factorial Approximate Maximum a Posteriori) algorithm [21] is used.

The unavailability of the appliance level consumption, for extracting the footprint, represents one of the main issues in the NILM supervised approach. In real scenarios, only the aggregated power consumption is available to the user. Therefore, the footprint extraction stage aims to extract the appliance footprint from the aggregated power: this work aims to investigate the performance of a footprint extraction procedure based on the HMM and AFAMAP algorithm.

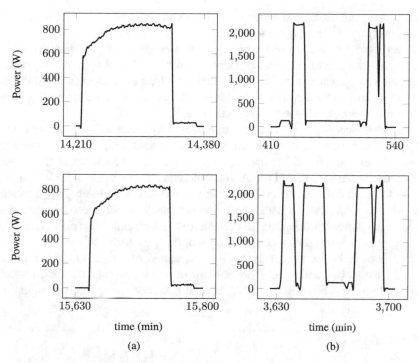

Fig. 4.21 Alike and different footprints for the same appliance, in ECO. (**a**) Dryer, household 1. (**b**) Dishwasher, household 2

A working cycle of an appliance is the interval between the power on and the power off by the user. In this time interval, the appliance power consumption signal is defined as *footprint*. Some examples of footprint taken from the ECO dataset [57] are shown in Fig. 4.21, that reports the power consumption traces recorded from the appliances located inside different Swiss households.

The usage of an appliance differs every time, especially in the case of equipments with different usage modes: e.g., the operating cycles of a washing machine can be set in a different way each time, or the operation of the dishwasher may vary according to the selected rinsing cycle. The different usage mode of the same appliance reflects on different footprint, as shown in Fig. 4.21b: the power levels in the two footprint of the dishwasher are the same, but they appear in different orders, which demonstrate that the working state comprising the appliance working cycle is unique, but they are employed in different orders, based on the user habits. Therefore, it is necessary to record different occurrence of the appliance footprint, in order to explore the different user habits in the appliance usage.

On the other hand, this aspect is not significant for appliances with easier working principle and a less complex circuit composition. In this case, the usage pattern of the appliance cannot be different in times, thus the footprint appears to be similar in each occurrence, as shown in Fig. 4.21a: the footprint of the dryer follows the same trend in time, which demonstrates the unique working cycle of the appliance and the unique way of usage by the user.

The footprint extraction is a necessary step in supervised NILM algorithms. In this context, the user exploits the aggregated power sensing system. An easy method to record the appliance footprint is to switch off all the appliances in the household and to turn on only the appliance of interest [15]. In this way, the aggregated power consumption corresponds to the appliance one.

The appliance switch on and off are detected by using a threshold schema on the active power consumption: when the value exceeds a threshold, the current is flowing in the circuit and the appliance is turned on, whereas when the value is below, the appliance is turned off. A threshold equal to the value of 50 W is a good choice for most datasets, nevertheless this value depends on the type of appliance and the activation power consumption. The samples between those two events are saved as the power consumption data related to the footprint. Multiple usages of the same appliance define different occurrences of the footprint.

In a household not all appliances can be turned off, e.g., the fridge and the freezer have to be continuously powered in order to maintain the food inside in safe condition. As shown in Fig. 4.22a, b, their power consumption is continuous in

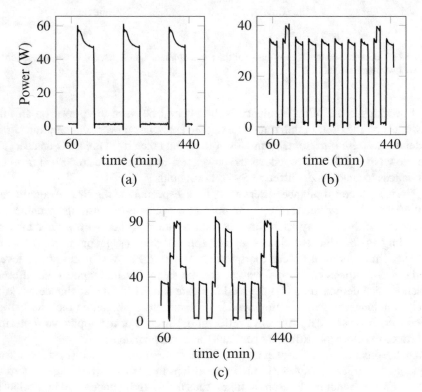

Fig. 4.22 Power consumption of continuously turned on appliances, in ECO. (**a**) Fridge, household 1. (**b**) Freezer, household 1. (**c**) Fridge-freezer combination, household 1

time, with a periodic working cycle. In this scenario, the aggregated consumption presents a continuous component, resulting from the sum of the fridge and freezer consumption, as shown in Fig. 4.22c. This signal can be modelled as the consumption of a unique model, representing the combination fridge-freezer as a composed appliance.

The presence of this component in the aggregated power does not allow to acquire a *clean* footprint of the appliance of interest, since all the appliances power signals are summed up on the aggregated power. Therefore, the footprint results to be *corrupted* and a procedure to clean it is needed.

In order to clean a corrupted footprint, a procedure to separate the fridge-freezer consumption from the appliance footprint one is needed.

The fridge-freezer contribution can be recorded on the aggregated power turning off all the other appliances in the household: in this way, the characterization of the fridge-freezer combination is not afflicted by noise or other appliances consumption, thus the extracted model results to be highly reliable and accurate.

The steps to be followed are the following:

1. the consumption of the fridge-freezer combination is recorded, in an adequate span of time to collect enough data for the modelling;
2. a corrupted version of the appliance of interest footprint is acquired;
3. the extraction procedure is applied to the recorded footprint, using the a priori knowledge of the fridge-freezer model and a generic model of the appliance.

The process of signal separation can be interpreted as a disaggregation problem with 2 sources: therefore, the same NILM algorithm, which is executed after the footprint extraction and the appliance modelling step, can be exploited for the footprint extraction step as well. In order to obtain the disaggregated traces, the NILM algorithm requires both the model of the fridge-freezer combination and of the appliance of interest. The first one is available, whereas the appliance model is not available, because the footprint extraction step precedes the appliance modelling step. Therefore, it is necessary to provide a generic model, which represents the class related to the appliance of interest, and which is suitably fitted on the specific appliance features, e.g., a priori knowledge of the maximum power consumption, in order to represent it as good as possible. This procedure introduces an uncertainty in the appliance modelling stage, which might be the cause of the error in the footprint extraction stage.

In this work, the NILM algorithm chosen for the disaggregation step is the *AFAMAP* proposed by *Kolter and Jaakkola* [21]: the algorithm requires the HMM of each appliance that contributes to the aggregated power signal.

From the analysis carried out in Sect. 4.4, the availability of the HMM of both the fridge-freezer combination and the appliance of interest is necessary. The first one is obtained from the corresponding consumption recorded, thus it is a model with high reliability: as showed in Fig. 4.22c, it is a model with 4 working states, derived from the composition of the 2 working states of the fridge and the freezer, whereas, for the appliance of interest, the model is not available, since it is derived after the footprint extraction step. Therefore, a generic HMM is exploited: it is obtained from a reference dataset, under the assumption that all the appliances of the same category act in the same way, while passing from a working state to another, so that the transition probability matrix results the same for each appliance in the category. Furthermore, it is assumed that the number of the working states is the same for all the appliances of the same category, since the working cycle of the appliance type observed in the footprint: therefore, the number of states is defined a priori for the appliance type, such as described in Table 4.3. In this approach, the univariate modelling case of the appliances consumption behaviour is considered.

For the appliances with a number of working states greater than 2, it is assumed that the consumption values are proportional to each other: therefore, the consumption values in the model are scaled based on the nominal (maximum) value, which is given a priori to the algorithm.

In this way, the HMM represents the appliance as good as possible, omitting the approximation on the consumption values of the middle working state and the approximation on the transition probability matrix.

After the AFAMAP algorithm execution, two disaggregated consumption profiles are obtained: the appliance one corresponds to the extracted footprint. Starting from this, the HMM representing the appliance is created, which is used in the disaggregation algorithm to solve the NILM problem.

In order to reach a good generalization in the HMM creation, the availability of different appliance footprints is necessary, as described in Sect. 4.4: this process allows to mitigate the errors introduced in the footprint extraction phase. A suggested value of occurrences to record is in the order of 10.

In Fig. 4.23 the flowchart of the footprint extraction algorithm is depicted. The diagram is composed of two sections: in the left one, the contribution of the fridge-freezer combination is recorded, from which the HMM is obtained; in the right one,

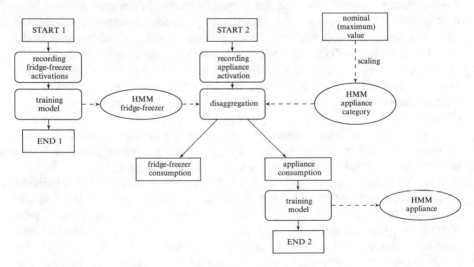

Fig. 4.23 Footprint extraction algorithm flowchart

the appliance activations are recorded, to obtain the footprint and the related HMM. This procedure is repeated for each appliance footprint recorded, which needs to be extracted.

4.4.1 Experimental Setup

The experiments have been conducted using different datasets: the first one for the generic model extraction, and the second one for testing the footprint extraction algorithm. The disaggregation experiments have been conducted on the same dataset, to evaluate the effectiveness of the footprint extraction algorithm, compared to the use of the true appliance level consumption, to create the appliance model.

The general model has been extracted using the *AMPds* dataset [58]. The experiments on footprint extraction and disaggregation are conducted on the *ECO* dataset [57], considering the households 1 and 2, whose appliances are:

- household 1: dryer, washing machine;
- household 2: dishwasher, oven.

The experiments include the fridge-freezer combination, present in each household.

4.4.2 Results

Figure 4.24 shows two examples of extracted footprints, compared to the original ones. In both cases, a good correspondence between the temporal trends can be noticed, which denotes that the model representing the fridge-freezer combination has a high reliability and it allows to extract the appliance footprint contribution in a suitable way. However, for several portions of the footprint, the correspondence with the power level is not correct: this might be due to the incorrect power levels of the general model, which are obtained from a scaling operation with respect to the nominal consumption value. Indeed, the error is introduced in the middle power levels, while for the maximum power level the correspondence is exact. In the entire process, the uncertainty introduced from the disaggregation algorithm, used to separate the footprint from the consumption of the fridge-freezer combination, needs to be considered.

The experiments have been conducted on a portion of 30 days of the ECO dataset. To evaluate the effectiveness of the footprint extraction procedure, the disaggregation results have been evaluated using:

- the models created by using the appliance level consumption, available in the dataset (*true* footprint);
- the models created by using the *extracted* footprint, following the procedure described in Sect. 4.4.

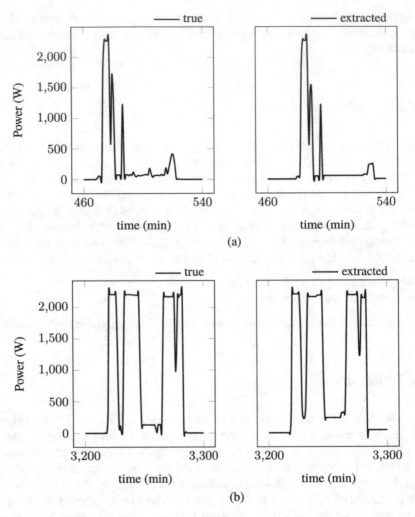

Fig. 4.24 Comparison between the true and the extracted footprint for some appliances. (**a**) Washing machine in ECO, household 1. (**b**) Dishwasher in ECO, household 2

The disaggregation results have been evaluated using the Precision (P) and Recall (R) metrics, defined in Sect. 2.4 in state and energy based sense. To compare the performance of the entire disaggregation system, the F-score (F_1) metric averaged across the appliances (Overall) has been used.

The parameters used in the AFAMAP algorithm were the same employed in Sect. 4.2. The disaggregation window parameter has been set $T = 60$ min.

The disaggregation results are showed in Tables 4.7 and 4.8. For both metrics, the algorithms achieve good performance: the best results are reached in the household 2 experiment, with an $F_1^{(S)}$ of 0.898 and $F_1^{(E)}$ of 0.956. This is due to the relatively simple problem studied in those cases: a disaggregation problem

Table 4.7 Disaggregation performance in ECO, household 1

Metric		Fridge-freezer	Dryer	Washing machine	Overall	Footprint
State based	$P^{(S)}$	0.506	0.657	0.909	0.691	True
	$R^{(S)}$	0.568	0.821	0.948	0.779	
	$F_1^{(S)}$	**0.536**	**0.730**	**0.928**	**0.732**	
	$P^{(S)}$	0.483	0.622	0.880	0.661	Extracted
	$R^{(S)}$	0.531	0.788	0.937	0.752	
	$F_1^{(S)}$	**0.506**	**0.695**	**0.908**	**0.704**	
Energy based	$P^{(E)}$	0.955	0.488	0.849	0.764	True
	$R^{(E)}$	0.815	0.972	0.978	0.922	
	$F_1^{(E)}$	**0.879**	**0.650**	**0.909**	**0.835**	
	$P^{(E)}$	0.953	0.422	0.809	0.728	Extracted
	$R^{(E)}$	0.790	0.976	0.982	0.916	
	$F_1^{(E)}$	**0.864**	**0.589**	**0.887**	**0.811**	

Bold values represent the higher performance in the algorithm comparison. Thus, the algorithm with bold values is the best algorithm in the experiment

Table 4.8 Disaggregation performance in ECO, household 2

Metric		Fridge-freezer	Dishwasher	Oven	Overall	Footprint
State based	$P^{(S)}$	0.741	0.926	0.977	0.881	True
	$R^{(S)}$	0.781	0.980	0.984	0.915	
	$F_1^{(S)}$	**0.760**	**0.952**	**0.980**	**0.898**	
	$P^{(S)}$	0.735	0.855	0.972	0.854	Extracted
	$R^{(S)}$	0.773	0.974	0.982	0.910	
	$F_1^{(S)}$	**0.754**	**0.911**	**0.977**	**0.881**	
Energy based	$P^{(E)}$	0.983	0.873	0.973	0.943	True
	$R^{(E)}$	0.944	0.983	0.984	0.970	
	$F_1^{(E)}$	**0.963**	**0.925**	**0.979**	**0.956**	
	$P^{(E)}$	0.981	0.816	0.975	0.924	Extracted
	$R^{(E)}$	0.939	0.982	0.988	0.970	
	$F_1^{(E)}$	**0.960**	**0.891**	**0.982**	**0.946**	

Bold values represent the higher performance in the algorithm comparison. Thus, the algorithm with bold values is the best algorithm in the experiment with only 3 appliances, with highly distinguishable values of power consumption, reveals to be solvable with high accuracy. The experiments in Table 4.8 show a better performance with respect to Table 4.7: the reason is the appliances footprints and the resulting HMMs composition. Indeed, the second problem is composed of models with a lower number of states (e.g., 3 states for the dishwasher, 3 states for the oven, with respect to the 3 states for the dryer and 4 states for the washing machine), thus the disaggregation problem results to be simpler in the resolution, and the overall performance reaches higher values. This trend was already introduced from the author of the disaggregation algorithm [21], who shows that the higher is the number of states related to the HMM, the higher is the complexity of the problem definition, and lower is the disaggregation performance due to the more difficult resolution. Regarding the first problem, the fridge-freezer combination has the consumption values close to the dryer ones, which leads to an ambiguity during the problem resolution and a lower performance for the total problem. In general, the appliance with the better performance is the one with the higher power consumption value: for the first problem the washing machine, for the second one the oven.

In both experiments the results corresponding to the true footprint show higher performance with respect to the extracted footprints ones: it means that the footprint extraction procedure introduces an error in the appliance modelling stage, which results in an error during the disaggregation algorithm resolution. Nevertheless, the results of the extracted footprint experiments show performance with an admissible relative loss: for the household 1 experiment, the relative loss results of 3.83% in state based sense, and 2.87% in energy based sense, while for the household 2 experiment, it results of 1.89% in state based sense, and 1.05% in energy based sense.

In conclusion, the models obtained after the footprint extraction procedure show a good correspondence with the original ones, which means that the footprint extraction is sufficiently reliable. Therefore, the footprint extraction algorithm introduced in this work provides a convenient procedure to the user for modelling the appliance at the cost of an acceptable loss in disaggregation performance.

Chapter 5
DNN Based Approach

Abstract The recent success of Deep Neural Networks (DNN) in several application scenarios drove the scientific community to employ this paradigm also for
NILM. Kelly and Knottenbelt compared three alternative DNNs: in the first, they
employed a convolutional layer followed by long short-term memory (LSTM)
layers to estimate the disaggregated signal from the aggregate one. In the second,
a denoising autoencoder composed of convolutional and fully connected layers is
trained to provide a denoised signal from the aggregate one. The third network
estimates the start time, the end time and the mean power demand of each appliance.
The algorithms were evaluated on the UK-DALE dataset and showed superior
performance with respect to the combinatorial optimization and FHMM algorithms
implemented in the Non-intrusive Load Monitoring Toolkit (NILMTK).

Keywords Deep neural network · Denoising autoencoder · Footprint · Active
power · Reactive power

5.1 Neural NILM

The work by Kelly and Knottenbelt [31] compared three different neural network
architectures: in the first, they employed a convolutional layer followed by LSTM
layers [60] to estimate the disaggregated signal from the aggregated one. In
the second, a denoising autoencoder (dAE) composed of convolutional and fully
connected layers is trained to provide a denoised signal from the aggregated one.
The third network estimates the start time the end time, and the mean power demand
of each appliance. The algorithms were evaluated on the UK-DALE dataset and the
results showed that the dAE approach outperforms the alternative neural networks
architectures as well as the FHMM algorithm implemented in the Non-intrusive
Load Monitoring Toolkit (NILMTK) [73].

5.2   Denoising AutoEncoder Approach

The NILM task can be formulated as a denoising problem by expressing the aggregated signal as the sum of the power consumption of the appliance of interest and a noise component that incorporates all the remaining contributions. In particular, Eq. (2.1) can be reformulated as:

$$\overline{y}(t) = y^{(j)}(t) + v^{(j)}(t), \tag{5.1}$$

for $j = 1, 2, \ldots, N$, where

$$v^{(j)}(t) = \sum_{\substack{i=1 \\ i \neq j}}^{N} y^{(i)}(t) + e(t), \tag{5.2}$$

represents an overall noise term for the appliance j that comprises both the measurement noise and the contributions of the other appliances. Thus, for obtaining $y^{(j)}(t)$, it would be sufficient to remove the noise term $v^{(j)}(t)$ from the aggregate measurement $\overline{y}(t)$.

In [31] and similarly in [30], noise removal is performed by means of a dAE, i.e., a neural network that is trained to reconstruct a clean signal from its noisy version presented at the input. Denoising autoencoders have been originally formulated in the context of *representation learning* and as an unsupervised training method [97]. The same structure has been later employed to perform actual noise removal, such as in speech related tasks [98, 99]. An autoencoder can be seen as an encoder network followed by a decoder network. The encoder provides an internal representation of the input signal and the decoder transforms it back into the input signal domain. A common choice consists in creating a network with specular encoder and decoder topologies. In the context of NILM, for each appliance, an autoencoder is trained to reconstruct the ground truth $y^{(j)}(t)$ given the aggregated signal $\overline{y}(t)$.

5.3   Algorithm Improvements

In this section, several algorithmic and architecture improvements to the dAE approach for NILM are proposed and an exhaustive comparative evaluation with the AFAMAP (Additive Factorial Approximate Maximum a Posteriori) algorithm [21] is conducted. In particular, compared to [31] the dAE approach for load disaggregation is improved by conducting a detailed study on the topology of the network, and by introducing pooling and upsampling hidden layers, and the rectifier linear unit (ReLU) activation function [100] in the output layer. Additionally, the network output is recombined by using a median filter on the overlapped portions of the disaggregated signal. The second contribution is an exhaustive performance comparison between AFAMAP and the dAE approach. Indeed, FHMMs have been largely employed in the last years since they are an effective approach for load disaggregation, and AFAMAP, in particular, received noteworthy attention by the scientific community [101, 102], as described in Sect. 4.1. However, an exhaustive performance comparison between the two methods has not been yet conducted. Indeed, the authors of [31] compare their proposed approaches to the FHMM method implemented in NILMTK [73], but their comparison does not consider more advanced FHMM algorithms such as AFAMAP [21]. Additionally, their experiments consider only a noised scenario on a single dataset (UK-DALE). Here, the evaluation is performed on three datasets, UK-DALE [61], AMPds [58] and REDD [29] in different conditions: firstly, the algorithms are evaluated on denoised and noised scenarios. In the denoised scenario, the aggregated signal is the sum of the power profiles of the appliances that are disaggregated. In the noised scenario, the aggregated signal comprises also measurement noise and the contributions of unknown appliances. Successively, the algorithms generalization capabilities are evaluated by performing disaggregation on the data acquired in a house not considered in the training phase (unseen scenario). The performance is evaluated by using both energy-based metrics and state-based metrics [73]: the first, evaluate the capability of the algorithm to estimate the actual power profile of the appliances, while the second the capability of estimating whether the appliance is in the "on" or "off" state. In order to perform the experiments in presence of noise, a Rest-of-the-World (RoW) model has been introduced in the original AFAMAP [21] algorithm. This model represents all the appliances but the ones of interest and makes AFAMAP able to operate in a noised scenario. The obtained results show that on average the dAE approach outperforms AFAMAP in all the addressed experimental conditions.

The general network topology proposed here for NILM is shown in Fig. 5.1: the encoder network (Fig. 5.1a) is composed of one or more one-dimensional convolutional layers that process the input signal and produce a set of feature maps. Each convolutional layer is followed by a linear activation function, by a max pooling layer, and by additional convolutional and pooling layers. Finally, one or more fully connected layers followed by a ReLU [100] activation function close the encoder network. The max pooling operation returns the maximum value within a neighbourhood, and in image processing, it makes the obtained representation invariant to small translations of the input. In NILM, this translates into being more independent on the location of an activation inside an analysis window. Additionally, max pooling reduces the size of the feature maps and the number of units in the fully connected layers, thus reducing the number of training parameters. The ReLU activation function calculates the maximum between its input and zero, and in this case it prevents the occurrence of negative values of the disaggregated active power. The decoder (Fig. 5.1b) is structured specularly to the encoder, with

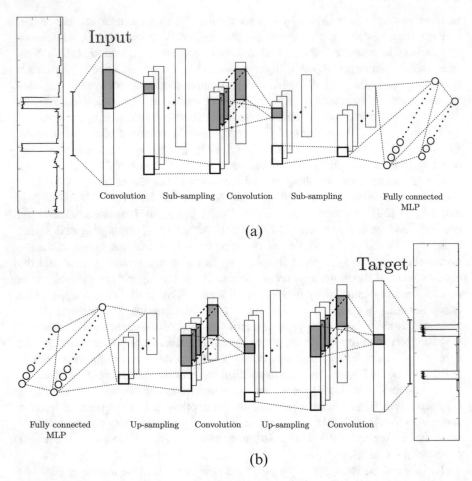

Fig. 5.1 Generic autoencoder architecture employed for disaggregation. (**a**) Encoder network. The input signal is the aggregated power consumption. (**b**) Decoder network. The target signal is ground truth power consumption of each appliance

upsampling layers taking the place of max pooling layers. Compared to [31], several network topologies are explored, with multiple convolutional stages, max pooling and upsampling layers are introduced, and the ReLU activation function in the fully connected layers.

The dAE network is trained to minimize the mean squared error between its output and the activation of a single appliance. Training is performed by using the Stochastic Gradient Descent (SGD) algorithm with Nesterov momentum [103], and with the early-stopping criterion to prevent overfitting. The input data and the target are normalized in order to improve the learning efficiency. With respect to the reference work [31], several advancements have been introduced in the training phase. In particular, during the training phase, the initial value of the learning rate is decreased when the performance on a validation set decreases. When this occurs, training is resumed from the epoch where the performance started decreasing. If the validation performance remains confined in a certain interval, typically when the learning process has reached the convergence or the learning rate has become too little, the early-stopping criterion is used. This is adopted in order to prevent overfitting.

In the disaggregation phase, the input signal $\overline{y}(t)$ is analysed by using sliding windows whose lengths depend on the size of the appliance activations. Windows are partially overlapped and the output signal is recombined by using a median filter on the overlapped portions. This differs from what proposed in [31], where the authors recompose the overlapped portions by calculating their mean value. The problem with this solution is that when an activation is only partially comprised in the analysis window, the network tends to underestimate the value of the output signal. As the window slides, the estimate increases, but averaging the overlapped portions produces an overall underestimated signal. Differently, by using the median operation on the overlapped portions, this phenomenon is mitigated, since greater values are preserved. The overall operation is depicted in Fig. 5.2.

The input signal is normalized following the same technique used in the training phase, while the disaggregated traces are denormalized after recombining outputs.

5.3.1   Experimental Setup

In order to conduct an exhaustive evaluation on different scenarios, three public datasets have been chosen. The Almanac of Minutely Power dataset (AMPds) [58] contains recordings of consumption profiles belonging to a single home in Canada for a period of 2 years, at 1 min sampling period. The experiments are conducted by using six appliances: dryer, washing machine, dishwasher, fridge, electric oven and heat pump. The second dataset, UK-DALE [61], is composed of consumption profiles recorded in five houses in UK over 2 years, at 6 s sampling period. The houses consumptions are not equally distributed over this time period, e.g., house 3 contains only the kettle consumptions and some minor appliances recordings, thus it is not considered in the experiments. The five target appliances considered in all the experiments are: fridge, washing machine, dish washer, kettle and microwave. The third dataset, REDD [29], contains aggregate and circuit-level power profiles of several US households. The sampling period of the aggregate data is 1 s, while the one of the target profiles is 3 s, thus aggregate data was downsampled in order to match the sample period of the target profiles. The experiments are conducted by using four appliances: dryer, dishwasher, fridge and microwave. In the seen scenario, the data from two houses is used both for training and testing. In the unseen scenario, the same data is used for training, while testing is performed on the data of a third house.

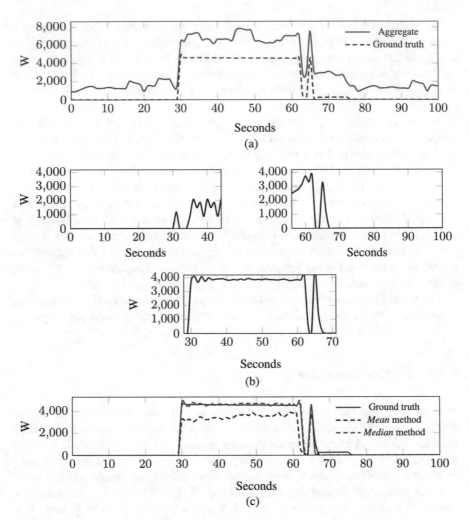

Fig. 5.2 Network outputs recombined by using the mean operation and the median operation recombination on the overlapped portions. (**a**) A portion of aggregated data, analysed with sliding window technique. (**b**) Output of the dAE for each window. (**c**) Disaggregated traces comparison between *median* and *mean* recombining methods

The chosen appliances represent the principal contributions to the peak of power consumption in the aggregated signal, which allows us to consider the *denoised* scenario as an approximation of the *noised* scenario in the traits of higher power consumption. On the other hand, the *noise* contribution, assigned to the RoW model, depends on the number of remaining appliances not modelled and on the total energy of the main aggregated signal, and this affects the disaggregation performance in the *noised* scenario. The *energy ratio* (ER), defined as:

$$\text{ER} = \frac{E_{\text{RoW}}}{E_{\text{main}}} = \frac{\sum_{t=1}^{T} e(t)}{\sum_{t=1}^{T} \overline{y}(t)}, \tag{5.3}$$

expresses the energy proportion between the RoW model and the total aggregated data, and the values for each house in the considered datasets are showed in Table 5.1.

The datasets are split in different portions for training and testing, and their dimensions depend on the availability of appliances activations within the dataset. Regarding the training procedure, within the period specified in Table 5.11, the first 20% of activations are used to compose the validation set, while the remaining 80% are used for the models training (Table 5.2).

Regarding the ground truth consumption availability, two different scenarios can be defined. In the *seen* scenario, the disaggregation is computed on the same houses used to train the models, but in different period from the training data. In this scenario, both models, HMM and neural network, are created exploiting the same portion of training, in order to conduct a fair comparison between the methods. On the other hand, in the *unseen* scenario, the disaggregation is computed on the data related to a house not considered in the training phase. In this scenario, the ground truth consumptions related to each appliance are not available in the house where the disaggregation is performed, therefore no training data can be considered to create the models. The generalization property of the neural network allows to avoid a training procedure and to use the model trained on a set of data different from the test, whereas the footprints need to be suitably extracted in order to train the HMM. One possible approach, in this sense, is represented by the user-aided

Table 5.1 Energy ratio (ER) for each house in the considered datasets

Dataset	AMPds	UK-DALE				REDD		
		House 1	House 2	House 4	House 5	House 1	House 2	House 3
ER	0.731	0.680	0.564	0.867	0.833	0.634	0.463	0.613

Table 5.2 Definition of the training, validation and test sets for the considered datasets

Dataset	Train+Validation	Test
AMPds	1 year, 6 months	6 months
UK-DALE		
House 1	1 year, 8 months, 3 days	7 days
House 2	4 months, 3 days	7 days
House 4	6 months, 25 days	7 days
House 5	2 months, 3 days	6 days
REDD		
House 1	33 days	3 days
House 2	12 days	2 days
House 3	12 days	6 days

footprint extraction algorithm, described in Sect. 4.4, that describes a procedure for the extraction of an approximated version of the appliance activations within the aggregated data when all the appliances are turned off, except the always-on in the house, i.e., the fridge and the freezer.

The experiments on the UK-DALE dataset have been performed as in [31], both for the *seen* and the *unseen* scenario.

The parameters related to the AFAMAP algorithm are defined as follows: the frame size is set to 60 min, which is an interval sufficiently large to include the whole activation for most of the appliances under study. For the ones with a longer activation, this frame size allows to include a complete operating subcycle, for which the HMM is still representative. The variance parameters are set to $\sigma_1^2 = \sigma_2^2 = 0.01$ according to the variance of the experimental data, and the regularization parameter is set to $\lambda = 1$. Table 5.3 presents the number of states, defined a-priori for each class of appliance. In the *denoised* scenario no parameters optimization has been conducted, whereas in the *noised* scenario, the number of the RoW states has been varied between the values {6, 8, 10} for both datasets.

The algorithm has been implemented in Matlab, and the CPLEX[1] solver has been adopted to solve the QP problem. The experiments have been conducted on a working station equipped with an Intel i7 CPU at 3.3 GHz, and 32 GB RAM. The time required for an experiments depends on the number of samples and the number of states of the HMM models: because of the different sampling rate between the datasets, the experiments last from 1 h for AMPds to 3 h for UK-DALE, while the introduction of the RoW model increases the simulation time up to 2 h for AMPds and 5 h for UK-DALE.

The parameters related to the dAE approach are defined as follows: each network receives data in a mini-batch of 64 sequences, and a mean and variance normalization is computed on the input data. In order to guarantee the same normalization over the whole dataset, the mean and variance values are computed from a random sample of the training set, whereas on the target data a min-max normalization is performed using the maximum power consumption value of the related appliance. The training data is composed of 50% of actual appliance related data, and 50% of synthetic data obtained by randomly combining real appliance activations. The training sequences have been extracted by using NILMTK [73]: this toolkit provides the method for the power activation extraction from the ground truth power consumption related to each appliance from both datasets. The data analysing window of the dAE needs to be enough large to comprise an entire activation of the

Table 5.3 Number of states m related to each class of appliance

Nr. of states	Dryer	Washing machine	Dishwasher	Fridge	Electric oven	Heat pump	Kettle	Microwave
m	3	4	3	2	3	3	2	2

---

[1]https://www-01.ibm.com/software/commerce/optimization/cplex-optimizer/.

Table 5.4 Window width (in samples) for the dAE architecture

Dataset	Dryer	Washing machine	Dishwasher	Fridge	Electric oven	Heat pump	Kettle	Microwave
UK-DALE	–	1024	1536	512	–	–	128	288
AMPds	75	120	210	45	120	90	–	–
REDD	1536	–	2304	496	–	–	–	96

The number of samples depends on the dataset sampling rate

appliance, but not too much to include other contributions, especially for appliances with short-duration activation. The window width depends on the appliance type, as described in Table 5.4.

As aforementioned, training has been performed by using the SGD algorithm with Nesterov momentum set 0.9. The maximum number of epochs has been set to 200 000, and the number of epochs for the variable step size technique has been set to 20 000. The initial value of the learning rate has been set to 0.1, with a decreasing factor equal to 10. The variable step size criterion has been applied on the $F_1^{(E)}$ calculated on the validation set, and the relative tolerance for early stopping criterion has been set equal to 0.01. The neural network has been implemented by means of the Lasagne library,[2] built on top of Theano [104]. All the network weights have been initialized randomly using Lasagne default initialization, without any layerwise pre training.

In [31], the network topology is composed of an input and an output convolutional layer with 8 kernels of size 4. The middle layers consist of 3 fully connected layers with ReLU activation functions, where the number of neurons in the central layer is equal to 128, whereas for the other layers the number depends on the length of the input sequence. In the disaggregation phase, a hop size of 16 samples has been considered. The performance of this work represents the baseline for this approach. An intensive parameters optimization has been conducted regarding the number of kernels (N), size of each kernel (S), and the number of neurons in the central layer (H). The experiments have been conducted using each combination of parameters within the ranges: N={2, 4, 8, 16, 32, 64}, S={2, 4, 8, 16, 32, 64}, H={8, 16, 32, 64, 128, 256, 512, 1024, 2048}. Kernels larger than the input size have not been considered. The architecture that achieves the highest performance has been used as a starting point of an additional campaign of experiment, for which the first convolutional layer has been preserved, and a second stage, including pooling and up-sampling layers, has been introduced. The parameters have been varied within the same ranges defined above.

Max pooling is calculated on a segment with sizes equal to 2 or 4 samples, and the overlapped portion is either equal to half of the window or not present. For this new architecture the experiments have been conducted with a full search of the optimal parameters. The disaggregation phase has been carried out with a sliding

---

[2]https://lasagne.readthedocs.io/en/latest/.

window technique over the aggregated signal, using overlapped window with hop size in the range $\{1, 2, 4, 8, \frac{1}{4}window, \frac{1}{2}window\}$, where *window* represents the window width defined in Table 5.4.

The number of networks tested for each appliance in three datasets has been varied from 150 to 200, and this experimental campaign has been conducted on both *denoised* and *noised* scenario, in the *seen* and *unseen* conditions.

The experiments have been conducted on nVIDIA K80 GPUs. The training time varies depending on the network dimension and appliance type: because of the different sampling rates of the datasets, the experiments require from 2 to 10 h depending on the size of the training set.

5.3.2 Results

Regarding the AFAMAP algorithm, in the *noised* scenario, preliminary experiments have demonstrated that the highest performance is obtained when the number of states of the RoW model is 6. For the sake of conciseness, only the results for that number of states are reported.

For the same reason, the results of the entire experimental campaign of the dAE algorithm will not be reported. For each scenario, the introduction of the second stage of CNN improves the performance with respect to the single CNN stage for the majority of appliances, as well as the effectiveness of the pooling layer. The experiments demonstrated that a hop size with 1 and 2 samples results in the best performance.

For the AMPds and UK-DALE datasets, the dAE algorithm outperforms AFAMAP both in the *noised* and the *denoised* scenarios, as shown in Tables 5.5, 5.6, Fig. 5.5a, b. More in details, Fig. 5.5 shows the radar charts related to the $F_1^{(E)}$ metric for each appliance, and the area inside a line gives an overall performance indicator of the related approach. On the AMPds dataset, in the *denoised* case study, the absolute improvement in terms of $F_1^{(E)}$ amounts to $+17.3\%$, while in the *noised* scenario the absolute improvement amounts to $+13.3\%$. The same trend can be observed by considering the other metrics. Compared to AFAMAP, NEP reduces by 2.012 in the *denoised* scenario, whereas it reduces by 3.819 in the *noised* scenario. State-based metrics show a similar trend, since, in the *denoised* case study, $F_1^{(S)}$ improves by $+24.7\%$, while in the *noised* case study the absolute improvement is $+29.8\%$. Similar remarks apply to MCC. Analysing the performance of the individual appliances, the dAE algorithm outperforms AFAMAP for all the appliances in both the *denoised* and the *noised* scenario. In terms of $F_1^{(E)}$, the highest absolute improvement can be observed for the dishwasher $(+45.9\%)$ in the *denoised* scenario, and for the oven in the *noised* scenario $(+48.4\%)$. Considering the other metrics, the dAE algorithm outperforms AFAMAP for all the appliances in both scenarios, except for the fridge in the *noised* scenario, where AFAMAP achieves lower NEP and higher $F_1^{(S)}$. Indeed, for this appliance in the

Table 5.5 Disaggregation performance in the seen scenario (AMPds dataset)

Scenario	Algorithm	Metric	Dryer	Washing machine	Dishwasher	Fridge	Electric oven	Heat pump	Overall
Denoised	AFAMAP [21]	$F_1^{(E)}$ (%)	87.3	14.5	44.4	35.5	38.1	76.9	60.4
		NEP	0.281	7.761	2.093	0.837	2.909	0.352	2.372
		$F_1^{(S)}$ (%)	60.7	7.4	11.9	36.0	5.0	86.2	50.3
		MCC	0.631	0.092	0.161	0.335	0.121	0.855	0.366
	dAE	$F_1^{(E)}$ (%)	96.1	50.5	90.3	63.7	84.1	77.4	**77.7**
		NEP	0.068	0.919	0.182	0.558	0.289	0.142	**0.360**
		$F_1^{(S)}$ (%)	76.0	54.8	76.8	75.6	53.4	93.4	**75.0**
		MCC	0.780	0.567	0.773	0.690	0.584	0.932	**0.721**
Noised	AFAMAP + RoW	$F_1^{(E)}$ (%)	65.3	6.2	27.8	38.3	8.9	54.6	40.8
		NEP	0.999	18.100	2.812	0.938	6.305	0.873	5.004
		$F_1^{(S)}$ (%)	16.0	7.6	10.7	43.3	2.0	55.5	30.8
		MCC	0.239	0.096	0.141	0.198	0.041	0.543	0.210
	dAE	$F_1^{(E)}$ (%)	91.2	11.9	49.8	39.1	57.3	65.4	**54.1**
		NEP	0.131	4.416	0.640	0.940	0.568	0.419	**1.185**
		$F_1^{(S)}$ (%)	76.8	10.8	58.2	33.1	45.9	79.8	**60.6**
		MCC	0.784	0.165	0.593	0.217	0.489	0.789	**0.506**

Numbers in bold indicate the best performing approach

Table 5.6 Disaggregation performance in the seen scenario (UK-DALE dataset)

Scenario	Algorithm	Metric	Kettle	Washing machine	Dishwasher	Fridge	Microwave	Overall
Denoised	AFAMAP [21]	$F_1^{(E)}$ (%)	93.4	64.3	48.1	79.1	84.1	77.4
		NEP	0.435	14.090	1.322	0.358	1.038	3.449
		$F_1^{(S)}$ (%)	81.9	41.2	22.5	84.6	78.1	70.4
		MCC	0.797	0.451	0.287	0.781	0.788	0.621
	dAE	$F_1^{(E)}$ (%)	94.1	59.6	86.2	85.8	82.9	**81.8**
		NEP	0.087	13.087	0.220	0.207	0.287	**2.777**
		$F_1^{(S)}$ (%)	95.7	56.2	57.4	93.2	90.4	**82.1**
		MCC	0.957	0.559	0.620	0.896	0.903	**0.787**
Noised	AFAMAP + RoW	$F_1^{(E)}$ (%)	12.8	20.4	18.5	49.4	11.5	24.9
		NEP	1.754	53.063	1.752	0.865	4.193	12.325
		$F_1^{(S)}$ (%)	7.79	15.80	16.95	51.91	18.24	35.49
		MCC	0.150	0.145	0.179	0.324	0.177	0.195
	Kelly [31]	$F_1^{(E)}$ (%)	80.1	35.1	58.2	64.1	59.5	60.4
		NEP	0.522	1.384	0.707	0.609	0.923	0.829
		$F_1^{(S)}$ (%)	82.12	35.32	69.53	65.68	62.58	69.18
		MCC	0.821	0.372	0.706	0.575	0.626	0.620
	dAE	$F_1^{(E)}$ (%)	82.4	54.8	84.3	73.6	72.4	**73.6**
		NEP	0.393	2.135	0.278	0.472	0.524	**0.760**
		$F_1^{(S)}$ (%)	86.6	40.8	55.6	78.2	75.5	**72.0**
		MCC	0.866	0.425	0.583	0.683	0.751	**0.661**

Numbers in bold indicate the best performing approach

noised scenario, the performance improvement in terms of $F_1^{(E)}$ is modest compared to the other appliances.

Compared to AFAMAP, in the UK-DALE dataset the absolute improvement in terms of $F_1^{(E)}$ is $+4.4\%$ in the *denoised* case study, and $+48.7\%$ in the *noised* scenario. The same trend can be observed by considering the other metrics: NEP reduces by 0.672 in the *denoised* scenario and by 11.564 in the *noised* scenario, while $F_1^{(S)}$ improves by $+11.7\%$ in the *denoised* case study and by $+36.51\%$ in the *noised* case study. MCC increases by 0.166 and by 0.466, respectively, in the *denoised* and in the *noised* scenario. Analysing the performance of the individual appliances, the dAE algorithm achieves superior performance for all the appliances in the *denoised* scenario, except for the washing machine and the microwave, for which the $F_1^{(E)}$ is similar. In the *noised* scenario, the dAE algorithm outperforms AFAMAP for all the appliances, with the highest improvement equal to $+69.6\%$ for the kettle. The same trend can be observed considering the other metrics. In the *noised* scenario, the optimization of the network parameters allows to outperform the dAE architecture presented in [31] for all the appliances, with the highest improvement of $F_1^{(E)}$ equal to $+26.1\%$ for the dishwasher. Considering the other metrics, the improvement follows the same trends, except for the washing machine evaluated in terms of NEP, and the dishwasher evaluated in terms of $F_1^{(S)}$ and MCC.

Regarding the REDD dataset (Table 5.7), in the *denoised* scenario the performance difference of the dAE algorithm with respect to AFAMAP varies with the evaluation metric. In particular, in terms of $F_1^{(E)}$ and MCC, AFAMAP outperforms the dAE algorithm, respectively, by 6.5% and 0.007. In terms of MCC, however, the relative improvement is limited, since it is equal to 0.95%. In terms of NEP and $F_1^{(S)}$, the dAE approach outperforms AFAMAP as shown in the experiments with the UK-DALE and AMPds datasets. This behaviour can be explained by considering that in the *denoised* seen scenario the HMM models in AFAMAP are trained by using data of the same building used in the disaggregation phase, while the network in the dAE approach is trained by using multiple buildings, and testing is performing on one of those. This aspect is less relevant in the *noised* scenario, because in AFAMAP the RoW model introduces a high variability in the disaggregation solution. Indeed, in this scenario the dAE approach outperforms AFAMAP regardless of the evaluation metric.

Generally, the dAE approach reaches higher disaggregation performance since it allows to reproduce complex activation profiles, which are learned during the training procedure and are associated to the aggregated profiles, even in the presence of the noise contribution. As shown in Tables 5.5, 5.6 and 5.7, the highest performance is reached in the disaggregation of the appliances with higher peak power consumption, since it allows a better association between the target and the aggregated input sequence during the training phase. In the HMM based approach, each state of an appliance model represents one value of power consumption, which does not allow to represent highly variable or transient phenomena between the working states of the appliance. Additionally, in the AFAMAP algorithm the disaggregation solution is obtained by considering all the appliance models at the

Table 5.7 Disaggregation performance in the seen scenario (REDD dataset)

Scenario	Algorithm	Metric	Dishwasher	Dryer	Fridge	Microwave	Overall
Denoised	AFAMAP [21]	$F_1^{(E)}$ (%)	67.1	94.4	80.5	80.2	**82.6**
		NEP	1.086	0.093	0.338	0.491	0.502
		$F_1^{(S)}$ (%)	50.12	97.59	89.79	66.86	78.85
		MCC	0.512	0.975	0.833	0.666	**0.746**
	dAE	$F_1^{(E)}$ (%)	66.3	87.3	74.6	64.9	76.1
		NEP	0.515	0.265	0.543	0.397	**0.430**
		$F_1^{(S)}$ (%)	71.7	92.7	80.5	69.6	**80.9**
		MCC	0.669	0.926	0.666	0.695	0.739
Noised	AFAMAP + RoW	$F_1^{(E)}$ (%)	36.4	31.9	36.0	17.9	35.4
		NEP	2.207	1.187	0.905	2.287	1.646
		$F_1^{(S)}$ (%)	32.9	57.6	39.6	16.2	46.0
		MCC	0.354	0.567	0.260	0.176	0.339
	dAE	$F_1^{(E)}$ (%)	62.1	72.5	66.9	53.0	**66.1**
		NEP	0.551	0.506	0.760	0.615	**0.608**
		$F_1^{(S)}$ (%)	64.0	81.8	70.9	61.6	**72.4**
		MCC	0.495	0.814	0.468	0.604	**0.595**

Numbers in bold indicate the best performing approach

same time, while in the dAE approach each network operates independently from the others. This may cause a false energy assignment to an appliance, due to the need to satisfy the constraint that the sum of the reconstructed profiles corresponds to the aggregated power. In presence of noise, the performance degrades significantly, since the presence of the RoW, composed of a higher number of states compared to appliance models, increases the number of admissible solutions and, as a consequence, the chance of errors in the disaggregated profiles reconstruction. Moreover, in the AFAMAP algorithm there is no information on the total duration of the complete activation, since appliance models incorporate only the information on the working state transition and on the consumption values.

Further evaluations can be carried out by analysing the disaggregated profiles in *denoised* and *noised* scenario. Considering the UK-DALE experiments in *seen* scenario, the profiles related to the dishwasher in the house 1 are shown in Fig. 5.3. The appliance activation is correctly detected by the dAE in both scenarios, without producing false positives in the disaggregated trace. In the *noised* scenario, the reconstructed profiles have a high uncertainty, caused by the presence of noise in the aggregated power, but the average energy in the activation has a good correspondence with the ground truth one, which demonstrates the low degradation of performance compared to the *denoised* scenario. The same experiment has been considered for the fridge, whose profiles are shown in Fig. 5.4. The dAE algorithm recognizes the appliance activation in the *denoised* scenario, with a less accurate profile reconstruction in the activation overlapped with other appliances with respect to the isolated ones. Differently, the performance degrades in the *noised* scenario, with an incorrect activation detection and the production of some false positives, caused by the presence of noise in the aggregated signal.

As aforementioned, the *unseen* scenario is evaluated by using the UK-DALE and REDD datasets, due to the availability of recordings from several houses in both.

As in the *noised seen* scenario, preliminary experiments conducted by varying the number of states in the RoW model demonstrated that the highest $F_1^{(E)}$ is obtained with 6 states. Similarly, for the dAE algorithm the results of the entire experimental campaign will not be reported for the sake of conciseness. For each scenario, the introduction of the second stage of CNN and of the pooling operation improves the performance with respect to the single CNN stage for the majority of the appliances. Regarding the hop size in the sliding window disaggregation phase, as in the *seen* scenario the highest performance is reached by using 1 and 2 samples.

Similarly to the *seen* scenario in the UK-DALE dataset, the baseline [31] performance for each appliance in the *noised* scenario is outperformed by means of the optimization of the network parameters, with the highest absolute improvement of $F_1^{(E)}$ equal to $+30.2\%$ for the washing machine. The same trend can be observed for the other metrics, excepting for the $F_1^{(S)}$ and the MCC, where the dishwasher performance degrades.

For both datasets, the dAE algorithm outperforms AFAMAP in both scenarios, as shown in Tables 5.9 and 5.8. In the UK-DALE dataset, the absolute improvement in terms of $F_1^{(E)}$ amounts to $+8.6\%$ in the *denoised* case study, whereas it increases to

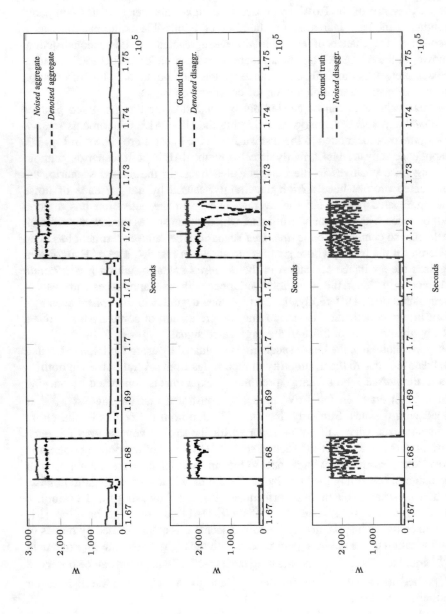

Fig. 5.3 Disaggregated profiles in *denoised* and *noised* scenario in UK-DALE dataset, *seen* case study, related to the dishwasher in house 1

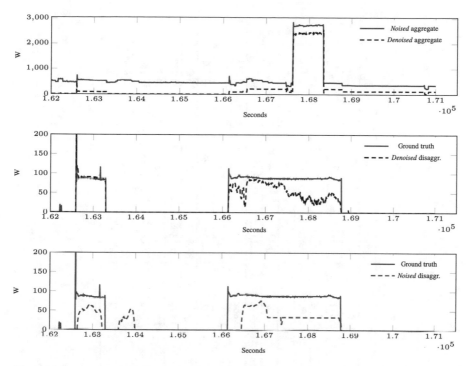

Fig. 5.4 Disaggregated profiles in *denoised* and *noised* scenario in UK-DALE dataset, *seen* case study, related to the fridge in house 1

+ 50.5% in the *noised* scenario, demonstrating the superiority of the neural network based approach with respect to the HMM one, especially in presence of the noise contribution. The results evaluated with the other metrics confirm the same trend, with a reduction of NEP equal to 0.543 in the *denoised* case study and to 5.418 in the *noised* case study. Considering the state based metrics, the improvement evaluated with the $F_1^{(S)}$ amounts to + 12.52% in the *denoised* scenario and + 53.10% in the *noised*, as well as regarding the MCC with an absolute improvement of + 0.170 in the *denoised* scenario and + 0.594 in the *noised* scenario. As showed in Fig. 5.5c, overall the dAE algorithm outperforms AFAMAP both in the *denoised* and in the *noised* scenarios. In particular, the dAE exhibits a noteworthy robustness against the presence of noise, while the $F_1^{(E)}$ of AFAMAP reduces significantly. Observing the results of each appliance, the highest absolute improvement is obtained for the kettle and it is equal to + 80.4%. In the *denoised* scenario, the dAE algorithm outperforms AFAMAP for all the appliances, with the only exception of the dishwasher where the $F_1^{(E)}$ is 1.6% lower. Considering the other metrics, in the *noised* scenario, the performance is improved for all the appliances, while in the *denoised* scenario the same trend can be observed, except for the washing machine, which degrades its performance in terms of NEP, $F_1^{(S)}$ and MCC.

Table 5.8 Disaggregation performance in the unseen scenario (REDD dataset)

Scenario	Algorithm	Metric	Dishwasher	Dryer	Fridge	Microwave	Overall
Denoised	AFAMAP [21]	$F_1^{(E)}$ (%)	32.2	49.5	69.3	7.0	46.4
		NEP	3.336	0.811	0.491	4.754	2.348
		$F_1^{(S)}$ (%)	18.6	89.8	73.6	4.3	55.9
		MCC	0.282	0.901	0.650	0.056	0.472
	dAE	$F_1^{(E)}$ (%)	76.1	83.7	78.5	60.5	**76.6**
		NEP	0.348	0.292	0.426	0.470	**0.384**
		$F_1^{(S)}$ (%)	87.5	85.8	88.1	67.4	**84.2**
		MCC	0.877	0.860	0.805	0.711	**0.813**
Noised	AFAMAP + RoW	$F_1^{(E)}$ (%)	22.5	18.8	40.0	8.4	26.4
		NEP	3.803	1.521	0.946	3.728	2.500
		$F_1^{(S)}$ (%)	14.2	41.3	37.3	5.1	35.0
		MCC	0.228	0.399	0.180	0.083	0.222
	dAE	$F_1^{(E)}$ (%)	41.8	57.2	60.4	13.6	**47.6**
		NEP	0.756	0.955	1.053	1.752	**1.129**
		$F_1^{(S)}$ (%)	49.2	59.3	71.7	16.8	**54.6**
		MCC	0.543	0.617	0.497	0.166	**0.456**

Numbers in bold indicate the best performing approach

Table 5.9 Disaggregation performance in the unseen scenario (UK-DALE dataset)

Scenario	Algorithm	Metric	Kettle	Washing machine	Dishwasher	Fridge	Microwave	Overall
Denoised	AFAMAP [21]	$F_1^{(E)}$ (%)	95.1	32.3	80.9	73.6	29.9	66.1
		NEP	0.114	2.089	0.457	0.449	3.311	1.284
		$F_1^{(S)}$ (%)	97.11	25.59	12.84	74.68	38.33	61.68
		MCC	0.971	0.353	0.177	0.690	0.440	0.526
	dAE	$F_1^{(E)}$ (%)	95.7	39.5	79.3	91.1	67.1	**74.7**
		NEP	0.056	2.406	0.371	0.195	0.675	**0.741**
		$F_1^{(S)}$ (%)	99.7	23.4	54.5	95.5	65.1	**74.2**
		MCC	0.997	0.286	0.604	0.931	0.664	**0.696**
Noised	AFAMAP + RoW	$F_1^{(E)}$ (%)	3.2	4.7	20.2	32.2	4.2	17.3
		NEP	3.087	6.559	2.078	1.021	18.413	6.231
		$F_1^{(S)}$ (%)	0	5.1	11.8	33.6	3.3	18.7
		MCC	−0.001	0.085	0.151	0.120	0.090	0.089
	Kelly [31]	$F_1^{(E)}$ (%)	79.1	23.3	39.2	65.1	20.6	50.8
		NEP	0.448	1.607	0.892	0.562	2.875	1.277
		$F_1^{(S)}$ (%)	93.9	26.5	55.9	77.8	30.9	66.8
		MCC	0.940	0.373	0.597	0.712	0.416	0.608
	dAE	$F_1^{(E)}$ (%)	83.6	53.5	69.2	78.7	45.8	**67.8**
		NEP	0.177	1.439	0.648	0.419	1.383	**0.813**
		$F_1^{(S)}$ (%)	95.6	67.5	50.9	82.8	45.4	**71.8**
		MCC	0.957	0.687	0.502	0.757	0.510	**0.683**

Numbers in bold indicate the best performing approach

On the REDD dataset, the absolute improvement in terms of $F_1^{(E)}$ amounts to $+30.20\%$ in the *denoised* scenario and $+21.18\%$ in the *noised* scenario. The other metrics follow the same trends, with a reduction of NEP equal to 1.964 in the *denoised* case study and to 1.371 in the *noised* case study. Considering the state based metrics, the improvement evaluated with the $F_1^{(S)}$ amounts to $+28.3\%$ in the *denoised* scenario and $+19.60\%$ in the *noised*, as well as regarding the MCC with an absolute improvement of $+0.341$ in the *denoised* scenario and $+0.234$ in the *noised* scenario. In the REDD dataset, differently from the *seen* scenario described above, the dAE algorithm outperforms on each appliance in both scenario, with the highest improvements in terms of $F_1^{(E)}$ of $+53.51\%$ for the microwave, except for the dryer in the *denoised* scenario with the state based metrics. The radar chart represented in Fig. 5.5e shows this improvement, and it represents the performance loss of both algorithm in the *noised* scenario with respect to the *denoised* scenario.

In the *unseen* scenario the generalization property of the dAE approach allows to apply the model without the need of training, with a reasonable degradation of performance. Regarding the AFAMAP algorithm, the approximation introduced by the footprint extraction procedure causes a lack of correspondence between the HMM and the appliance working states consumptions, and this results in a higher performance degradation, particularly in presence of *noise* where RoW model is present. This demonstrates the effectiveness of the neural networks approaches in an *unseen* scenario, which is the most interesting condition, because it represents a real-world application of the NILM service. As described in the previous section, the state based metrics confirm that the dAE produces a more reliable activation detection, with respect to the HMM based approach, even in an unseen scenario.

5.4   Exploitation of the Reactive Power

Besides machine learning techniques employed in order to solve the NILM problem, in the literature, neural networks (NNs) have been widely explored to address the problem of NILM.

Reactive power has already been identified as an exploitable feature to enhance NILM performances: starting from the seminal work of Hart [15], where the appliances working states are detected in the complex plan exploiting the active and reactive power consumption, up to the use of reactive power to train transient-state models [47, 50], In [105], the authors propose an active learning approach to significantly reduce the number of training samples needed to achieve high classification accuracies. In [106], the authors include reactive power trajectories, on top of which a principal component analyser is built to model each appliance. Finally, in [107] a recent approach based on finite-state machine modelling is built on top of real and reactive power signatures.

In this work the problem of NILM is addressed by using a particular family of NNs, that is the convolutional autoencoder. In particular, following the formalization

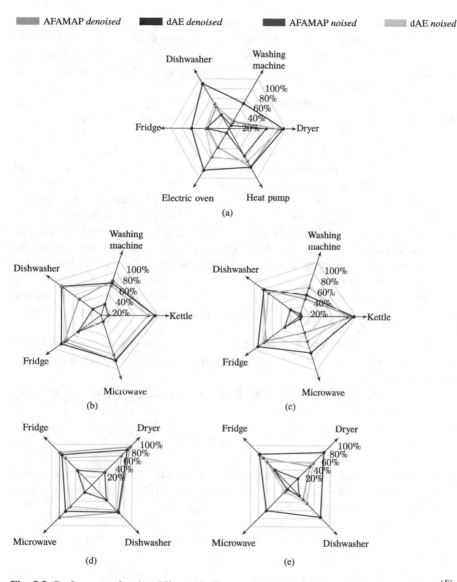

Fig. 5.5 Performance for the different appliances for all the addressed algorithms. The $F_1^{(E)}$ (%) is represented. (**a**) Disaggregation performance on the AMPds dataset, *seen* scenario. (**b**) Disaggregation performance on the UK-DALE dataset, *seen* scenario. (**c**) Disaggregation performance on the UK-DALE dataset, *unseen* scenario. (**d**) Disaggregation performance on the REDD dataset, *seen* scenario. (**e**) Disaggregation performance on the REDD dataset, *unseen* scenario

of the problem as a denoising case study, the analysed architecture will be named *denoising autoencoder* (dAE) hereafter. As described in Sect. 5.2, denoising autoencoders architectures were deeply explored and several advancements were introduced, demonstrating that this approach reaches higher performance with respect to the FHMM based one [21].

In the majority of the methods discussed above, the signal under analysis in the disaggregation algorithm is represented by the active aggregate power consumption. The main focus of this work is the analysis on how the reactive power aggregate signal, used as input feature, influences the performance of dAEs. To do so, dAEs have been trained in an asymmetrical configuration, where the input consisted of both active and reactive aggregate power signals, and the output was solely the active power appliance trace. The proposed approach has been evaluated on two publicly available datasets, the *Almanac of Minutely Power* dataset (AMPds) [58] and the *UK Domestic Appliance-Level Electricity* (UK-DALE) [61] dataset. Despite not all appliances seem to benefit from the introduction of the reactive power feature, the overall averaged scores show significant improvements in all considered scenarios.

In the present examination, NILM has been formalized and treated as a denoising problem. In this scenario, the aggregate active power is seen as the superimposition of the most relevant appliance consumptions, plus a rest-of-the-world noise term, as described by the (2.1).

This equation highlights that, for each appliance, it is possible to retrieve the corresponding active power consumption $y_a^{(j)}(t)$ by removing the noise term from the whole aggregate signal.

The denoising problem stated above allows us to look at the dAE as a mapping function f so that:

$$f : \mathbb{R}^{(L,1)} \Rightarrow \mathbb{R}^{(L,1)}, \tag{5.4}$$

where L is the signal's window length. This means that the denoising function f takes as argument a one-dimensional signal (the aggregate data) and retrieves, again, a one-dimensional vector: the disaggregated signal.

In introducing the reactive power signal, the active and reactive signals are concatenated on the second dimension, therefore the mapping function f will now follows:

$$f : \mathbb{R}^{(L,2)} \Rightarrow \mathbb{R}^{(L,1)}. \tag{5.5}$$

This solution considers that the dAE will be driven to exploit the correlation existing between active and reactive consumptions. Invariance on the third axis is imposed by using one-dimension convolutional kernels, thus the active and reactive power signals are treated similarly to different colour channels in image processing tasks with convolutional neural networks.

Other configurations are possible, such as the concatenation on the time axis or the use of two separated dAE chains. These settings, however, would introduce discontinuities in the input data, or, in the latter case, they would increase the computational complexity of the method. Due to these considerations, the proposed setting appeared as the best choice to make use of convolutional layers and to avoid excessive complications of the network topology.

In the present study, data has been pre-processed similarly to [31], where also the multiple values of L (with reference to Eqs. 5.4 and 5.5) are given for each appliance. Hereafter, only a short description of the most salient pre-processing steps is given.

Firstly, for each appliance, active time windows are identified and grouped under the name of *activations*. An activation is defined as a time window in which the appliance consumption exceeds a *minimum active power* threshold for more than a *minimum ON time*. Moreover, if a subsequent consumption peak occurs before a pre-set *minimum OFF time*, it will be placed inside the same activation. Finally, all time windows between two adjacent activations will be grouped as *inactive sections*.

In constructing the target of an active sequence, one activation is randomly extracted and shifted: this way the network will be shown multiple perspectives of the same activation, making the most out of its informative potential. On the other hand, inactive sequence targets will be synthesized as zero-numbered vectors: its associated input will be an aggregate time window picked from the inactive section ensemble. The inactive input window will also be randomly shifted.

Finally, both the active and reactive components are standardized by subtracting the sequence mean value from each sample, and dividing it by the standard deviation calculated over the entire dataset:

$$\tilde{y}_c(t) = \frac{\overline{y}_c(t) - \overline{y}_{c,\text{mean}}}{\overline{y}_{c,\text{std}}}, \tag{5.6}$$

where $\overline{y}_{c,\text{mean}}$ is the active $(c = a)$ and reactive $(c = r)$ sequence mean value, calculated on L samples, and $\overline{y}_{c,\text{std}}$ is the global standard deviation of the active and reactive signals. As mentioned in [31], this independent sequence centring does lose information, but it is able to improve the generalization capabilities of the network. Target sequences, on the other hand, are simply divided by the maximum power value of each appliance:

$$\underline{y}_a^{(i)}(t) = \frac{y_a^{(i)}(t)}{\max y_a^{(i)}(1:T)}, \tag{5.7}$$

where $\max y_a^{(i)}(1:T)$ is, for each appliance, the maximum power indicated in [31] and used in the activation extraction phase.

The dAE topology can be divided into two main stages: an encoder and a decoder. The first dAE stage, the encoder, takes as input the aggregate signal. The input is firstly processed by convolutional and pooling layers to extract shift-invariant features; then, fully-connected layers are used to extract higher-level feature representations. In the encoding phase, max-pooling is used as sub-sampling function. The convolutional layers are composed exploiting linear activation function, whereas the fully-connected encoding layers are composed exploiting rectifier linear unit (ReLU). In this book, the encoder is composed of two convolutional layers and one fully-connected layer. The number of kernels and their size, the dimension of the max-pooling window and the number of units of the fully-connected layer have been explored in the experimental phase.

The encoder's output is fed to the decoder, whose hyper-parameter configuration and topology mirror the structure of the encoding network. Therefore, the decoder's input is firstly processed by fully-connected layers, followed by convolutional and up-sampling ones. The only noticeable difference between the encoder and the decoder topologies resides in the activation function: in the decoder, the rectifier function is used in place of the linear one. We remind that, despite generally being a symmetrical structure, the decoder output is always a uni-dimensional vector, meaning that, even when the reactive power is used as input, the output is always trained to match only the disaggregated active power signal.

Networks are trained with a supervised approach, aiming, for each input time window, to minimize the mean squared error between the disaggregated output and the (measured) corresponding appliance consumption. In order to minimize the mean squared error loss, the stochastic gradient descent algorithm is used, with the addition of Nesterov momentum [108] to further speed up the training convergence.

During training, networks are also shown synthetic sequences of data. The synthesis procedure is the same as described in [31], and it consists in randomly summing appliance activations with random shifts so to generate synthetic aggregate data. In addition to generating synthetic sequences, the algorithm will also make sure that active and inactive sequences will be used with a 50-50 ratio.

In order to prevent overfitting and excessive training times, an early-stopping criterion is used. However, in evaluating the model performance, the model's energy-based F1 score is used in place of the mean squared measure. Every time the model performance is checked on a validation set, the algorithm evaluates if an improvement has been made over the registered best score. If no improvements are encountered for a fixed number of training iterations, the difference between the last score and the best one is calculated; if such difference is higher than a certain threshold the learning rate is reduced and the training is re-started from the last best-performing configuration, otherwise the training is stopped. With such approach we aim at avoiding to stop the training when big score fluctuations occur (possibly) because the training cost function has not yet reached a stable minimum.

In the disaggregation phase, the whole aggregate signal is processed by the trained dAEs, which, for each appliance, reconstructs the corresponding consumption. The processing takes place with a sliding window approach, where overlapping sequences of fixed size are shown to the network, and the respective outputs are collected. In order to re-combine the overlapping sequences a median filtering has been used, since in Sect. 5.3 it was found to perform better than the average recombination used in [31].

Finally, at the end of the disaggregation phase, all samples are up-scaled by the same maximum power factor max $y_a^{(i)}(1 : T)$ previously used to scale the network targets.

5.4.1   Experimental Setup

AMPds contains recordings taken in a single house from 21 different power meters, with a sampling period of 60 s. The time period covered consists of 2 years, going from April 1, 2012 to March 31, 2014. Additionally to the active power consumption, AMPds also contains apparent and reactive power signals for the whole measurement period.

The UK-DALE dataset contains measurements taken in five different houses at multiple sampling rates. Differently from AMPds, in UK-DALE only active and apparent power measurements have been recorded, and this does not apply to all houses: in house three and four the aggregate active power signal was not measured. Therefore, in this evaluation only data taken from house one, two and five is used, with the sampling period set to 6 s.

Despite no reactive power measurements are available for the UK-DALE dataset, in order to retrieve the needed reactive power aggregate signals, the following relationship has been used:

$$\overline{y}_r(t) = \sqrt{\left(\overline{y}_{ap}(t)\right)^2 - \left(\overline{y}_a(t)\right)^2}, \tag{5.8}$$

where $\overline{y}_r(t)$, $\overline{y}_{ap}(t)$ and $\overline{y}_a(t)$ represent the reactive, the apparent and the active power sample in each sequence, respectively. On the AMPds, on the other hand, reactive power measurements allowed us to evaluate the magnitude of its contribution over the active power, at appliance level consumption. In particular, as shown in Table 5.10, reactive over active signal ratios were calculated for each appliance. What emerges is that the reactive power's magnitude oscillates between 7.6% and 26.6% of the active one, thus highlighting that, in this scenario, reactive power can indeed be considered a significant additional feature.

As shown in Table 5.11, data has been divided, for both datasets, into training, validation and test sets. In particular, after training activations are extracted, 20% of

Table 5.10 Reactive over active (R/A) power ratios on the AMPds

Appliance	R/A (%)
Dishwasher	7.6
Electric oven	10.3
Fridge	9.5
Heat pump	20.0
Tumble dryer	16.1
Washing machine	26.6

Table 5.11 Training, validation and test data subdivision by start and end date

Dataset	Number of buildings	Train+Validation set	Test set
AMPds	1	2012-10-01	2012-04-01
		2014-04-01	2012-10-01
UK-DALE	1	2013-04-12	2014-10-22
		2014-10-21	2014-12-15
	2	2013-05-22	2013-09-27
		2013-09-26	2013-10-10
	5	2014-06-29	2014-09-01
		2014-09-01	2014-09-07

Table 5.12 Seen and unseen building subdivision for the UK-DALE dataset

Appliance	Train/seen test	Unseen test
Dishwasher	1, 2	5
Fridge	1, 2	5
Kettle	1, 2	5
Microwave	1, 2	5
Washing machine	1, 5	2

them is used to form validation batches, and the remaining 80% is used to construct train batches.

For each appliance trained on the UK-DALE dataset, data from one of the three available houses is excluded from training. This allows to define two different test cases, namely the *seen* and the *unseen* scenarios. By doing so, the aim is to test more deeply the networks' ability to generalize, since, in the unseen scenario, the model is not given the possibility to overfit the corresponding appliance signal. In Table 5.12 the seen/unseen house subdivision is reported.

Here a description of the parameter setup used in our experiments is given. Firstly, the window sizes used for each appliance are the same as shown in Sect. 5.3. These window sizes were identified as best-performing by Kelly and Knottenbelt in [31]; also all thresholds used during the activation extraction phase are the same as indicated in Kelly's article.

At the beginning of the training phase, network parameters are initialized with a random distribution: control is taken over this and all other random processes via the pre-setting of the code's random seed. The network training is conducted batch-by-batch, with a batch size fixed to 64 sequences, and the same size is used also for the validation batch. The maximum number of training iterations is fixed to 200,000, and the validation check is performed once every 10 iterations. We choose 2000 to be the maximum number of no-improvement iterations: when reached, the algorithm will decide whether to stop the training or to reduce the learning rate by a factor of 10.

In order to evaluate different network topologies, a grid search has been conducted on the encoder hyperparameters. All the combinations of the following hyperparameters have been evaluated:

- first layer kernels: 32, 128;
- kernel window size: 4, 16, 32;
- pooling size: 2, 4;
- fully-connected layer size: 512, 4096.

In addition, the number of kernels in the second convolutional layer is double than the first layer ones, and the pooling and window sizes are equal in both convolutional layers. As aforementioned, the topology of the decoder mirrors the one of the encoders. Considering each appliance and the two datasets, the total number of experiments run is 528. Moreover, it has to be highlighted that, given the absence of reactive and apparent power measurements at appliance level for the UK-DALE, the data synthesis procedure described in Sect. 5.3 has not been possible. Therefore, only appliances trained on the AMPds made use of the described data augmentation technique. The activation extraction procedure explained in Sect. 5.3 has been performed by using the Non-Intrusive Load Monitoring Toolkit (NILMTK) [73]. The experimental framework (available upon request) has been developed in Python (v. 2.7.10) and Keras (v. 2.1.2) over the Theano [104] backend (v. 0.9.0). Finally, the hardware setup used to run our experiments were NVIDIA GeForce GTX 970 and TITAN X graphic processing units.

5.4.2 Results

In Tables 5.14 and 5.13 experimental results obtained on both datasets are reported. Moreover, to better visualize score trends, the reader can refer to Fig. 5.6, where radar graphic representations of the scores are showed.

Observing AMPds results, it is possible to notice that all appliances benefit from the introduction of the reactive power input, the only exception being the electric oven. By looking at the overall scores, it is possible to identify an improvement of 8.1% if both the active and reactive aggregate signals are used instead of the active-only input.

Table 5.13 F-score results (%) on the UK-DALE dataset

Appliance	Seen		Unseen	
	Active	Active + Reactive	Active	Active + Reactive
Dishwasher	71.6	**83.3**	44.3	**50.6**
Fridge	68.5	**70.8**	68.9	**76.7**
Kettle	89.0	**89.9**	**82.1**	80.2
Microwave	64.4	**80.7**	37.0	**67.9**
Washing machine	35.5	**49.8**	5.4	**23.3**
Overall	67.7	**76.1**	58.0	**62.9**

Bold values represent the experiment with the higher result, therefore the best algorithm in the experiment comparison

Table 5.14 F-score results
(%) on the AMPds

Appliance	Active	Active + Reactive
Dishwasher	62.9	**77.5**
Electric oven	**66.3**	65.0
Fridge	37.4	**43.4**
Heat pump	72.7	**76.3**
Tumble dryer	94.8	**95.5**
Washing machine	34.3	**59.5**
Overall	62.1	**70.2**

Bold values represent the experiment with the
higher result, therefore the best algorithm in the
experiment comparison

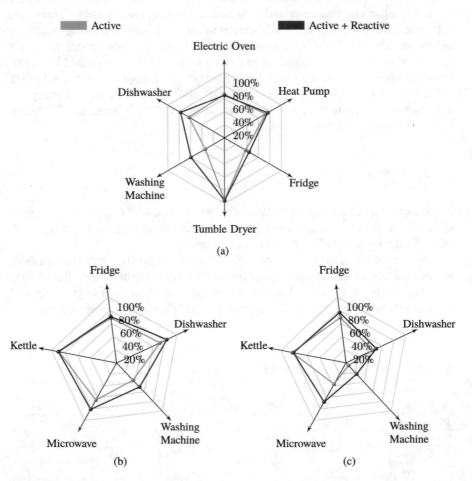

Fig. 5.6 Performance for the different appliances for the all the addressed algorithms. The
$F_1^{(E)}$ (%) is represented. (**a**) Disaggregation performance on the AMPds dataset, *seen* scenario.
(**b**) Disaggregation performance on the UK-DALE dataset, *seen* scenario. (**c**) Disaggregation
performance on the UK-DALE dataset, *unseen* scenario

Concerning the UK-DALE dataset, improvements can be observed on both the seen and the unseen scenarios. In particular, for the seen scenario, all appliances show score improvements ranging from 0.9% for the kettle, to 16.3% for the microwave, resulting in an overall improvement of 8.4%. On the unseen scenario, on the other hand, only the kettle showed reduced performance with the introduction of reactive power: −1.9%. The overall score, however, improves by 4.9%, highlighting that score increases still outweigh the reduction encountered.

Finally, it is worth highlighting that, given the nature of the proposed algorithm, an ensemble technique can be employed. As there is no dependence among appliance models, it is possible to choose the best-performing input configuration, namely active power only or active and reactive power. A possible strategy for choosing whether to use the reactive power as additional feature is by observing the F1 scores obtained in the validation phase. On the AMPds dataset, this solution translates into using the active power only model for the electric oven, and the active and reactive power models for the remaining appliances, resulting in a 0.2% improvement. This also shows that the validation score generally gives a reliable information on which configuration can be preferred. The ensemble technique has no effect on the UK-DALE dataset, since the validation scores obtained by using the active and reactive power models are always higher than the scores obtained with active power only models.

Chapter 6
Conclusions

Abstract In this book, the Machine Learning approaches for Non-Intrusive Load Monitoring have been studied. Within all the techniques explored by the scientific community, this work has been focused on the hidden Markov model based and the deep neural network based, since their capability and promising performance at the forefront of the improvements could be introduced.

Keywords Conclusion · Future works · Performance improvement · Gaussian mixture models · Neural rest-of-the-world model

In this book, the Machine Learning approaches for Non-Intrusive Load Monitoring have been studied. Within all the techniques explored by the scientific community, this work has been focused on the hidden Markov model based and the deep neural network based, since their capability and promising performance at the forefront of the improvements could be introduced.

For the HMM based approaches, firstly the appliance modelling and all the related aspects have been introduced, therefore the AFAMAP algorithm and its method improvements have been described. Specifically, the variation on the formulation has been detailed for the exploitation of the reactive power. The algorithm has been tested on both denoised and noised scenario, by means of the usage of the Rest-of-the-world model. The last aspect discussed deals with a facilitated procedure for the footprint extraction related to a specific appliance from the aggregated data.

For the DNN based approaches, the dAE has been introduced and the optimization in the model training phase and in architecture has been described. In addition, the recombining technique in the disaggregation phase has been improved. The algorithm has been tested in both denoised and noised scenario, for which a different optimization procedure has been conducted, as well as in the case of seen and unseen scenario. As last aspect, the exploitation of the reactive power has been considered in the network architecture, providing its own optimization procedure in all the considered scenario.

In Chap. 2, an updated review of the state of the art regarding the NILM algorithms is presented, together with an updated list of available datasets, which are typically used for parameter tuning and evaluation purposes. For what concern the NILM methods addressed in this review, they were first divided into two main categories: load classification and source separation algorithms. This reflects the nature of the method for the disaggregation and the limits or the improvements which could be explored, despite the same problem statement. It is pointed out that most of the contributions make use of the sole active power signal, and only few methods use the reactive power (or the phase difference between the voltage/current phasors). Exploiting this information can be beneficial to obtain a performing disaggregation action, but, on the other hand, requires a specific hardware able to provide the needed measurements. Clearly, a direct comparison between all methods presented is not immediately possible, due to the difference in terms of performance criteria and involved datasets. Indeed, the metrics used in those works could vary, representing different aspects of the obtained results. In terms of performance, the most promising methods appeared to be the HMM models, which are widely used for their capability to represent the appliance consumption behaviour with a relatively easy training procedure. On the other hand, the DNN based method have been recently emplyed in NILM with promising performance, following the recent success in various Computational Intelligence fields.

In Chap. 3 the Machine Learning approaches, adopted in NILM methods, have been described.

The AFAMAP algorithm [21], resulting one of the most performing and computationally efficient among the HMM based approaches, has been described in Chap. 4. The appliance models based on HMM have been introduced and the procedure for estimating their parameters has been described. This consists in the extraction of the footprint of the appliance by means of an Appliance Activity Detector and in the estimation of the power levels of each working state by clustering the appliance footprint with the k-means algorithm. The same procedure is used to compose the Rest-of-the-World model for the testing of the FHMM algorithms in the noised scenario. AFAMAP is revised in order to improve its performance through a more exhaustive exploitation of the information pertaining to the appliance activity. The proposed algorithm exploits both additive and differential FHMM to model the activity of the appliances. At each time step, the best combination of appliances working state is chosen to represent the actual aggregated consumption: as a result of the optimization process, a set of coefficients are returned to weight the appliances working state and compose the own disaggregated consumption. The revised algorithm, however, takes into account additional elements. In regard to the FHMM model, a forward differential model is paired to the reference backward differential FHMM, thus not only the transition from the previous state to the current one is included, but also the transition from the current state to the next one. In addition, the use of solver boundaries is explored: firstly, the setting has been related to the admissible state combination of the aggregated power; alternatively, the reactive power disaggregation output has been used to select the boundaries, endorsing the heterogeneous data usage effectiveness. Later, active and reactive power have been introduced in Additive Factorial HMM for non-intrusive load monitoring. The disaggregation algorithm is an alternative formulation of the AFAMAP developed in order to deal with the bivariate formulation of the problem. As a result, the algorithm is able to output the disaggregated profiles in the active and reactive power components. The proposed approach has been compared to the univariate formulation of AFAMAP and to the algorithm presented by Hart in [15]. The latter is based on Finite State Machine appliance models and it employs both the active and reactive power. The algorithm has been improved for handling the occurrence of multiple solutions by means of a MAP technique. The experiments have been conducted on the AMPds [58] dataset, which provides the ground truth appliance consumption both in the active and reactive power components. The results showed that, in a denoised scenario, the proposed approach outperforms both the comparative methods, with an absolute F_1-Measure improvement of $+14.9\%$ and $+2.5\%$ in the 6 appliances case study. As last aspect, a footprint extraction procedure has been introduced as a solution for the appliance modelling in real NILM scenarios. Indeed, in order to create the appliance model and to use this in the disaggregation algorithm, the user needs to record the appliance consumption profile. A facilitated procedure is needed, in order to obtain a clean footprint from the aggregated power signal in real scenario: therefore, a user-aided footprint extraction procedure is defined. The solution introduced here relies on the availability of a general model for the appliance category to obtain the clean footprint. This is the starting point of the modelling stage: in this work the AFAMAP algorithm has been used. The resulting models have been tested in a disaggregation problem, and they have been compared with the same problem solved using the true appliance model, i.e., the models created using the actual footprint from the appliance level consumption. The results have showed a moderate performance reduction compared to the ideal case due to the footprint extraction stage. For those reasons, the footprint extraction procedure introduced in this work can be considered as an effective method for the user employment in a real NILM scenario.

In Chap. 5, a DNN architecture based on the denoising autoencoder topology has been proposed. Compared to the work by Kelly and Knottenbelt [31] several improvements have been introduced. In the training phase, the variable step size has been adopted, with an early stopping criterion based on the performance metric calculated on the validation set. In the disaggregation phase, the median filter has been applied to combine the overlapped portion of signal in the sliding window analysis of the aggregated power data. In order to achieve the best performance, for each network an optimization of the network parameters has been conducted, starting from the reference architecture and introducing a second layer of CNN and a pooling stage to compress the size of the output. The proposed approach has been compared to the AFAMAP [21] algorithm. This algorithm has been adopted for the *noised* scenario with the introduction of RoW model. The experiments have been conducted on the AMPds [58], on the UK-DALE [61] and on the REDD [29] datasets, evaluating both the *denoised* and *noised* scenario. Furthermore, the availability of recordings from more than one building in the UK-DALE and in the REDD datasets allowed to evaluate the algorithms on an *unseen* scenario. The results showed that the proposed approach outperforms the comparative methods in the overall average between the appliance, both in *denoised* and *noised* scenario. Regarding the *unseen* scenario, the performance demonstrated that the generalization property of the dAE allowed acceptable degradation of performance, with respect to the AFAMAP algorithm, in which the footprint extraction stage introduced errors in the HMM modelling phase.

Moreover, the use of reactive aggregate power has been analysed for enhancing dAE performances. A grid search has been conducted on different dAE hyperparameters, and networks have been evaluated on the AMPds [58] and on the UK-DALE [61] and in different scenarios. Improvements on the overall $F_1^{(E)}$ have been registered in both datasets, namely +8.1% on the AMPds, and +8.4% and 4.9% on the UK-DALE dataset. This allows us to conclude that the reactive power indeed provides significant information for NILM with dAEs.

6.1   Future Research Topics

Since the high interest regarding the consumption reduction and the recent improvement in the smart grid researches, the interest on improving those method will be certainly maintained, pointing out as good results as a distributed network of smart plug. For this reason, future works will be oriented on different aspect, related to each algorithm discussed above.

Regarding the AFAMAP algorithm, a more reliable appliance model will be considered in order to improve the representation of the working states, e.g., the usage of Gaussian Mixture Model (GMM) within the HMM allows the representation of a more suitable power level density distribution with respect to a simple Gaussian distribution. Furthermore, additional information about the working states duration will be introduced, allowing the discrimination of HMMs with similar transition probabilities but different time in the switching activity. This translates into a fully exploitation of the differential model. Additionally, an observation window of longer duration could be introduced in the differential model.

Regarding the appliance modelling stage, an unsupervised clustering technique will be introduced to automatically detect the number of power levels, e.g., regarding appliances which do not belong to the categories considered. Regarding the disaggregation and solver algorithms, binary variables will be introduced in the problem formalization, leading to a Mixed Integer Quadratic Program (MIQP), in order to impose the variable to assume binary results and not integer values as in fuzzy logic, which can lead to ambiguous evaluation in the HMM state evolution. Finally, further experiments will be conducted in order to compare the proposed solution to other approaches recently presented in the literature [26, 27, 35].

Regarding the user-aided footprint extraction procedure, the separation of the model representing the fridge-freezer combination in the single component will be evaluated, since the AFAMAP algorithm shows a better working in the problem resolution using models with lower number of states. Moreover, more experiments will be performed using different datasets in literature, in which a more detailed study about the generalization performance can be carried out, specially for the generic model selection.

For the dAE approach, the introduction of a constraint between the neural model output will be considered, in order to assume the equality between the aggregated data and the sum of the profiles reconstructed, in the *denoised* scenario. In order to apply this constraint in the *noised* scenario, the introduction of the neural based RoW model will be required.

Regarding the exploitation of the reactive power in the dAE algorithm, future works will be focused on the investigation regarding the appliance which degrades performance. Moreover, the reactive power consumption as target for each architecture will be considered, in order to allow the benefit from correlations between the active and reactive target signals.

Bibliography

1. D. Archer, *Global Warming: Understanding the Forecast*, 2nd edn. (Wiley, Hoboken, 2012)
2. C. Rosenzweig, D. Karoly, M. Vicarelli, P. Neofotis, Q. Wu, G. Casassa, A. Menzel, T.L. Root, N. Estrella, B. Seguin, P. Tryjanowski, C. Liu, S. Rawlins, A. Imeson, Attributing physical and biological impacts to anthropogenic climate change. Nature **453**(7193), 353–357 (2008)
3. N. Oreskes, The scientific consensus on climate change. Science **306**(5702), 1686–1686 (2004)
4. H. Farhangi, The path of the smart grid. IEEE Power Energy Mag. **8**(1), 18–28 (2010)
5. K. Carrie Armel, A. Gupta, G. Shrimali, A. Albert, Is disaggregation the holy grail of energy efficiency? The case of electricity. Energy Policy **52**, 213–234 (2013)
6. C. Fischer, Feedback on household electricity consumption: a tool for saving energy? Energy Effic. **1**(1), 79–104 (2008)
7. S. Darby, The effectiveness of feedback on energy consumption. Technical Report, University of Oxford, Oxford (2006)
8. K. Ehrhardt-Martinez, K.A. Donnelly, J.A. Laitner, Advanced metering initiatives and residential feedback programs: a meta-review for household electricity-saving opportunities. Technical Report E105, American Council for an Energy-Efficient Economy Washington, DC (2010)
9. J. Laitner, K. Ehrhardt-Martinez, V. McKinney, Examining the scale of the behaviour energy efficiency continuum, in *American Council for an Energy Efficient Economy, European Council for an Energy Efficient Economy Conference*, Cote d'Azur (2009), paper ID 1367
10. G.T. Gardner, P.C. Stern, The short-list: the most effective actions us households can take to curb climate change. Environ. Sci. Policy Sustain. Environ. **50**(5), 12–24 (2008)
11. S. Ahmadi-Karvigh, B. Becerik-Gerber, L. Soibelman, A framework for allocating personalized appliance-level disaggregated electricity consumption to daily activities. Energy Build. **111**, 337–350 (2016)
12. I. Abubakar, S.N. Khalid, M.W. Mustafa, H. Shareef, M. Mustapha, Application of load monitoring in appliances' energy management – a review. Renew. Sustain. Energy Rev. **67**, 235–245 (2017)
13. A. Zoha, A. Gluhak, M.A. Imran, S. Rajasegarar, Non-intrusive load monitoring approaches for disaggregated energy sensing: a survey. Sensors **12**(12), 16838–16866 (2012)
14. Z. Wang, G. Zheng, Residential appliances identification and monitoring by a nonintrusive method. IEEE Trans. Smart Grid **3**(1), 80–92 (2012)
15. G.W. Hart, Nonintrusive appliance load monitoring. Proc. IEEE **80**(12), 1870–1891 (1992)

© The Author(s), under exclusive license to Springer Nature Switzerland AG 2020   127
R. Bonfigli, S. Squartini, *Machine Learning Approaches*
to Non-Intrusive Load Monitoring, SpringerBriefs in Energy,
https://doi.org/10.1007/978-3-030-30782-0

16. M. Zeifman, K. Roth, Nonintrusive appliance load monitoring: review and outlook. IEEE Trans. Consum. Electron. **57**(1), 76–84 (2011)

17. H. Kim, M. Marwah, M.F. Arlitt, G. Lyon, J. Han, Unsupervised disaggregation of low frequency power measurements, in *Proceedings of the 11th SIAM International Conference on Data Mining*, Mesa (2011), pp. 747–758

18. O. Parson, S. Ghosh, M. Weal, A. Rogers, An unsupervised training method for non-intrusive appliance load monitoring. Artif. Intell. **217**, 1–19 (2014)

19. O. Parson, M. Weal, A. Rogers, A scalable non-intrusive load monitoring system for fridge-freezer energy efficiency estimation, in *Proceedings of the 2nd International Workshop on Non-Intrusive Load Monitoring*, Austin, Jun. 3 2014

20. A. Zoha, A. Gluhak, M. Nati, M.A. Imran, Low-power appliance monitoring using Factorial Hidden Markov Models, in *Proceedings of the 8th International Conference on Intelligent Sensors, Sensor Networks and Information Processing: Sensing the Future (ISSNIP)*, Melbourne, vol. 1 (2013), pp. 527–532

21. J.Z. Kolter, T. Jaakkola, Approximate inference in additive factorial HMMs with application to energy disaggregation. J. Mach. Learn. Res. **22**, 1472–1482 (2012)

22. I. Valera, F.J.R. Ruiz, F. Perez-Cruz, Infinite factorial unbounded-state hidden Markov model. IEEE Trans. Pattern Anal. Mach. Intell. **38**(9), 1816–1828 (2016)

23. Y. Li, Z. Peng, J. Huang, Z. Zhang, J.H. Son, Energy disaggregation via hierarchical factorial HMM, in *Proceedings of the 2nd International Workshop on Non-Intrusive Load Monitoring*, Austin (2014)

24. M. Zhong, N. Goddard, C. Sutton, Signal aggregate constraints in additive factorial HMMs with application to energy disaggregation, in *Advances in Neural Information Processing Systems* (2014), pp. 3590–3598

25. M. Zhong, N. Goddard, C. Sutton, Interleaved factorial non-homogeneous hidden Markov models for energy disaggregation, in *Proceedings of Advances in Neural Information Processing System, Workshop on Machine Learning for Sustainability*, Lake Tahoe (2014), pp. 1–5

26. A. Cominola, M. Giuliani, D. Piga, A. Castelletti, A.E. Rizzoli, A hybrid signature-based iterative disaggregation algorithm for non-intrusive load monitoring. Appl. Energy **185**(Part 1), 331–344 (2017)

27. S. Makonin, F. Popowich, I.V. Bajić, B. Gill, L. Bartram, Exploiting HMM sparsity to perform online real-time nonintrusive load monitoring. IEEE Trans. Smart Grid **7**(6), 2575–2584 (2016)

28. M.J. Johnson, A.S. Willsky, Bayesian nonparametric hidden semi-Markov models. J. Mach. Learn. Res. **14**(1), 673–701 (2013)

29. J.Z. Kolter, M.J. Johnson, REDD: a public data set for energy disaggregation research, in *Proceedings of the SustKDD Workshop on Data Mining Applications in Sustainability*, San Diego (2011), pp. 1–6

30. F.C.C. Garcia, C.M.C. Creayla, E.Q.B. Macabebe, Development of an intelligent system for smart home energy disaggregation using stacked denoising autoencoders, in *Proceedings of the International Symposium on Robotics and Intelligent Sensors (IRIS)*, Tokyo, Dec. 17–20 2016, pp. 248–255

31. J. Kelly, W. Knottenbelt, Neural NILM: deep neural networks applied to energy disaggregation, in *Proceedings of the 2nd ACM International Conference on Embedded Systems for Energy-Efficient Built Environments*, BuildSys '15 (ACM, New York, 2015), pp. 55–64

32. L. Mauch, B. Yang, A new approach for supervised power disaggregation by using a deep recurrent LSTM network, in *Proceedings of GlobalSIP*, Orlando (2015), pp. 63–67

33. L. Mauch, B. Yang, A novel DNN-HMM-based approach for extracting single loads from aggregate power signals, in *Proceedings of ICASSP*, Shanghai (2016), pp. 2384–2388

34. M.-S. Tsai, Y.-H. Lin, Modern development of an adaptive non-intrusive appliance load monitoring system in electricity energy conservation. Appl. Energy **96**, 55–73 (2012)

35. B. Zhao, L. Stankovic, V. Stankovic, On a training-less solution for non-intrusive appliance load monitoring using graph signal processing. IEEE Access **4**, 1784–1799 (2016)

36. M. Figueiredo, A. De Almeida, B. Ribeiro, Home electrical signal disaggregation for non-intrusive load monitoring (NILM) systems. Neurocomputing **96**, 66–73 (2012)
37. J.M. Gillis, S.M. Alshareef, W.G. Morsi, Nonintrusive load monitoring using wavelet design and machine learning. IEEE Trans. Smart Grid **7**(1), 320–328 (2016)
38. S.R. Shaw, S.B. Leeb, L.K. Norford, R.W. Cox, Nonintrusive load monitoring and diagnostics in power systems. IEEE Trans. Instrum. Meas. **57**(7), 1445–1454 (2008)
39. J. Froehlich, E. Larson, S. Gupta, G. Cohn, M. Reynolds, S. Patel, Disaggregated end-use energy sensing for the smart grid. IEEE Pervasive Comput. **10**(1), 28–39 (2011)
40. H.K. Alfares, M. Nazeeruddin, Electric load forecasting: literature survey and classification of methods. Int. J. Syst. Sci. **33**(1), 23–34 (2002)
41. B. Neenan, J. Robinson, R.N. Boisvert, *Residential Electricity Use Feedback: A Research Synthesis and Economic Framework* (Electric Power Research Institute, Palo Alto, 2009)
42. M. Severini, S. Squartini, F. Piazza, Hybrid soft computing algorithmic framework for smart home energy management. Soft Comput. **17**(11), 1983–2005 (2013)
43. M. Severini, S. Squartini, F. Piazza, Computational framework based on task and resource scheduling for micro grid design, in *2014 International Joint Conference on Neural Networks (IJCNN)*, July 2014, pp. 1695–1702
44. M.B. Figueiredo, B. Ribeiro, A. de Almeida, Electrical signal source separation via nonnegative tensor factorization using on site measurements in a smart home. IEEE Trans. Instrum. Meas. **63**(2), 364–373 (2014)
45. L. De Baets, J. Ruyssinck, C. Develder, T. Dhaene, D. Deschrijver, On the Bayesian optimization and robustness of event detection methods in NILM. Energy Build. **145**, 57–66 (2017)
46. L.K. Norford, S.B. Leeb, Non-intrusive electrical load monitoring in commercial buildings based on steady-state and transient load-detection algorithms. Energy Build. **24**(1), 51–64 (1996)
47. M. Berges, E. Goldman, H.S. Matthews, L. Soibelman, K. Anderson, User-centered nonintrusive electricity load monitoring for residential buildings. J. Comput. Civ. Eng. **25**(6), 471–480 (2011)
48. H.-H. Chang, H.-T. Yang, Applying a non-intrusive energy-management system to economic dispatch for a cogeneration system and power utility. Appl. Energy **86**(11), 2335–2343 (2009)
49. H.-H. Chang, Non-intrusive demand monitoring and load identification for energy management systems based on transient feature analyses. Energies **5**(11), 4569–4589 (2012)
50. H.-H. Chang, P.W. Wiratha, N. Chen, A non-intrusive load monitoring system using an embedded system for applications to unbalanced residential distribution systems. Energy Procedia **61**, 146–150 (2014)
51. M. Aiad, P.H. Lee, Unsupervised approach for load disaggregation with devices interactions. Energy Build. **116**, 96–103 (2016)
52. Y.F. Wong, Y.A. Sekercioglu, T. Drummond, V.S. Wong, Recent approaches to non-intrusive load monitoring techniques in residential settings, in *Proceedings of the IEEE Symposium on Computational Intelligence Applications In Smart Grid (CIASG)*, Singapore, 16–19 Apr. 2013, pp. 73–79
53. J.M. Alcalá, J. Ureña, Á. Hernández, D. Gualda, Sustainable homecare monitoring system by sensing electricity data. IEEE Sens. J. **17**(23), 7741–7749 (2017)
54. S.M. Tabatabaei, S. Dick, W. Xu, Toward non-intrusive load monitoring via multi-label classification. IEEE Trans. Smart Grid **8**(1), 26–40 (2017)
55. R.S. Butner, D.J. Reid, M. Hoffman, G.P. Sullivan, J. Blanchard, *Non-Intrusive Load Monitoring Assessment: Literature Review and Laboratory Protocol* (Pacific Northwest National Laboratory, Richland, 2013)
56. N. Batra, J. Kelly, O. Parson, H. Dutta, W. Knottenbelt, A. Rogers, A. Singh, M. Srivastava, NILMTK: an open source toolkit for non-intrusive load monitoring, in *Proceedings of the 5th International Conference on Future Energy Systems* (ACM, New York, 2014), pp. 265–276

57. C. Beckel, W. Kleiminger, R. Cicchetti, T. Staake, S. Santini, The ECO data set and the performance of non-intrusive load monitoring algorithms, in *Proceedings of the 1st ACM International Conference on Embedded Systems for Energy-Efficient Buildings (BuildSys 2014)*, Memphis. November 2014 (ACM, New York, 2014), pp. 80–89

58. S. Makonin, F. Popowich, L. Bartram, B. Gill, I.V. Bajic, AMPds: a public dataset for load disaggregation and eco-feedback research, in *Proceedings of the IEEE Electrical Power and Energy Conference (EPEC)*, Halifax (2013)

59. F. Paradiso, F. Paganelli, D. Giuli, S. Capobianco, Context-based energy disaggregation in smart homes. Future Internet **8**(1), 4 (2016)

60. S. Hochreiter, J. Schmidhuber, Long short-term memory. Neural Comput. **9**, 1735–1780 (1997)

61. J. Kelly, W. Knottenbelt, The UK-DALE dataset, domestic appliance-level electricity demand and whole-house demand from five UK homes. Scientific Data **2**, 150007 (2015)

62. H. Shao, M. Marwah, N. Ramakrishnan, A temporal motif mining approach to unsupervised energy disaggregation: applications to residential and commercial buildings, in *Proceedings of the Twenty-Seventh AAAI Conference on Artificial Intelligence* (AAAI Press, Palo Alto, 2013), pp. 1327–1333

63. K. Anderson, A. Ocneanu, D. Benitez, D. Carlson, A. Rowe, M. Berges, BLUED: a fully labeled public dataset for event-based non-intrusive load monitoring research, in *Proceedings of the 2nd KDD Workshop on Data Mining Applications in Sustainability (SustKDD)*, Beijing, August 2012

64. S. Barker, A. Mishra, D. Irwin, E. Cecchet, P. Shenoy, J. Albrecht, Smart*: an open data set and tools for enabling research in sustainable homes, in *SustKDD*, August 2012

65. A. Reinhardt, P. Baumann, D. Burgstahler, M. Hollick, H. Chonov, M. Werner, R. Steinmetz, On the accuracy of appliance identification based on distributed load metering data, in *Proceedings of the 2nd IFIP Conference on Sustainable Internet and ICT for Sustainability (SustainIT)*, October 2012

66. C. Holcomb, Pecan street inc.: a test-bed for NILM, in *International Workshop on Non-Intrusive Load Monitoring*, Pittsburgh (2012)

67. J.-P. Zimmermann, M. Evans, J. Griggs, N. King, L. Harding, P. Roberts, C. Evans, Household electricity survey: a study of domestic electrical product usage, Intertek Testing & Certification Ltd (2012)

68. N. Batra, M. Gulati, A. Singh, M.B. Srivastava, It's different: insights into home energy consumption in India, in *Proceedings of the 5th ACM Workshop on Embedded Systems for Energy-Efficient Buildings* (ACM, New York, 2013), pp. 1–8

69. A. Monacchi, D. Egarter, W. Elmenreich, S. D'Alessandro, A.M. Tonello, GREEND: an energy consumption dataset of households in Italy and Austria, CoRR, vol. abs/1405.3100 (2014)

70. N. Batra, O. Parson, M. Berges, A. Singh, A. Rogers, A comparison of non-intrusive load monitoring methods for commercial and residential buildings, CoRR, vol. abs/1408.6595 (2014)

71. M. Maasoumy, B. Sanandaji, K. Poolla, A.S. Vincentelli, BERDS - Berkeley energy disaggregation data set, in *Proceedings of the Workshop on Big Learning at the Conference on Neural Information Processing Systems (NIPS 2013)* (2014)

72. L. Pereira, F. Quintal, R. Gonçalves, N.J. Nunes, SustData: a public dataset for ICT4S electric energy research, in *ICT for Sustainability 2014 (ICT4S-14)* (Atlantis Press, Amsterdam, 2014)

73. J. Kelly, N. Batra, O. Parson, H. Dutta, W. Knottenbelt, A. Rogers, A. Singh, M. Srivastava, NILMTK v0.2: a non-intrusive load monitoring toolkit for large scale data sets: demo abstract, in *Proceedings of the 1st ACM Conference on Embedded Systems for Energy-Efficient Buildings*, New York, BuildSys '14 (ACM, New York, 2014), pp. 182–183

74. B.W. Matthews, Comparison of the predicted and observed secondary structure of T4 phage lysozyme. Biochim. Biophys. Acta (BBA) Protein Struct. **405**(2), 442–451 (1975)

75. C. Beleites, R. Salzer, V. Sergo, Validation of soft classification models using partial class memberships: an extended concept of sensitivity and co. applied to grading of astrocytoma tissues. Chemom. Intell. Lab. Syst. **122**, 12–22 (2013)

76. L. Rabiner, A tutorial on hidden Markov models and selected applications in speech recognition. Proc. IEEE **77**(2), 257–286 (1989)

77. Z. Ghahramani, M.I. Jordan, Factorial Hidden Markov models. Mach. Learn. **29**(2–3), 245–273 (1997)

78. S. Haykin, *Adaptive Filter Theory*, 3rd edn. (Prentice-Hall, Upper Saddle River, 1996)

79. D.E. Rumelhart, G.E. Hinton, R.J. Williams, Learning representations by back-propagating errors. Nature **323**, 533–536 (1986)

80. Y. Lecun, L. Bottou, Y. Bengio, P. Haffner, Gradient-based learning applied to document recognition. Proc. IEEE **86**(11), 2278–2324 (1998)

81. I. Goodfellow, Y. Bengio, A. Courville, *Deep Learning* (MIT Press, Cambridge, 2016). http://www.deeplearningbook.org

82. I. Goodfellow, H. Lee, Q.V. Le, A. Saxe, A.Y. Ng, Measuring invariances in deep networks, in *Advances in Neural Information Processing Systems*, vol. 22, ed. by Y. Bengio, D. Schuurmans, J.D. Lafferty, C.K.I. Williams, A. Culotta (Curran Associates, Red Hook, 2009), pp. 646–654

83. Y. Bengio, P. Lamblin, D. Popovici, H. Larochelle, Greedy layer-wise training of deep networks, in *Advances in Neural Information Processing Systems*, vol. 19, ed. by B. Schölkopf, J.C. Platt, T. Hoffman (MIT Press, Cambridge, 2007), pp. 153–160

84. P. Vincent, H. Larochelle, I. Lajoie, Y. Bengio, P.-A. Manzagol, Stacked denoising autoencoders: learning useful representations in a deep network with a local denoising criterion. J. Mach. Learn. Res. **11**, 3371–3408 (2010)

85. S. Pattem, Unsupervised disaggregation for non-intrusive load monitoring, in *2012 11th International Conference on Machine Learning and Applications (ICMLA)*, vol. 2 (IEEE, Piscataway, 2012), pp. 515–520

86. O. Parson, S. Ghosh, M. Weal, A. Rogers, Non-intrusive load monitoring using prior models of general appliance types, in *Twenty-Sixth Conference on Artificial Intelligence (AAAI-12)* (2012)

87. S. Makonin, F. Popowich, Nonintrusive load monitoring (NILM) performance evaluation. Energy Effic. **8**(4), 809–814 (2014)

88. J.A. Hartigan, M.A. Wong, Algorithm as 136: a k-means clustering algorithm. J. R. Stat. Soc. Ser. C Appl. Stat. **28**(1), 100–108 (1979)

89. D. Egarter, M. Pöchacker, W. Elmenreich, Complexity of power draws for load disaggregation, CoRR (2015)

90. A.I. Cole, A. Albicki, Algorithm for nonintrusive identification of residential appliances, in *Proceedings of the IEEE International Symposium on Circuits and Systems (ISCAS)*, Monterey, 31 May - 3 Jun. 1998, pp. 338–341

91. C. Laughman, K. Lee, R. Cox, S. Shaw, S. Leeb, L. Norford, P. Armstrong, Power signature analysis. IEEE Power Energy Mag. **1**(2), 56–63 (2003)

92. A. Marchiori, D. Hakkarinen, Q. Han, L. Earle, Circuit-level load monitoring for household energy management. IEEE Pervasive Comput. **10**(1), 40–48 (2011)

93. M. Weiss, A. Helfenstein, F. Mattern, T. Staake, Leveraging smart meter data to recognize home appliances, in *Proceedings of IEEE International Conference on Pervasive Computing and Communications (PerCom)*, Lugano (2012), pp. 190–197

94. H. Goncalves, A. Ocneanu, M. Berges, Unsupervised disaggregation of appliances using aggregated consumption data, in *Proceedings of the 1st KDD Workshop on Data Mining Applications in Sustainability (SustKDD)*, San Diego (2011)

95. A. Krizhevsky, Learning multiple layers of features from tiny images, M.S. thesis, University of Toronto, 2009

96. G. Hinton, L. Deng, D. Yu, G.E. Dahl, A.-r.Mohamed, N. Jaitly, A. Senior, V. Vanhoucke, P. Nguyen, T.N, Sainath, et al., Deep neural networks for acoustic modeling in speech recognition: the shared views of four research groups. IEEE Signal Process. Mag. **29**(6), 82–97 (2012)

97. P. Vincent, H. Larochelle, I. Lajoie, Y. Bengio, P.-A. Manzagol, Stacked denoising autoencoders: learning useful representations in a deep network with a local denoising criterion. J. Mach. Learn. Res. **11**(3), 3371–3408 (2010)

98. S. Araki, T. Hayashi, M. Delcroix, M. Fujimoto, K. Takeda, T. Nakatani, Exploring multi-channel features for denoising-autoencoder-based speech enhancement, in *Proceedings of the IEEE International Conference on Acoustics, Speech and Signal Processing*, Brisbane, 19–24 Apr. 2015, pp. 116–120

99. X. Lu, Y. Tsao, S. Matsuda, C. Hori, Speech enhancement based on deep denoising autoencoder, in *Proceedings of Interspeech*, Lyon, 25–29 Aug. 2013, pp. 436–440

100. V. Nair, G.E. Hinton, Rectified linear units improve restricted Boltzmann machines, in *Proceedings of the 27th International Conference on Machine Learning (ICML)*, Haifa, 21–24 Jun. 2010, pp. 807–814

101. A. Gabaldon, R. Molina, A. Marín-Parra, S. Valero-Verdu, C. Alvarez, Residential end-uses disaggregation and demand response evaluation using integral transforms. J. Mod. Power Syst. Clean Energy **5**(1), 91–104 (2017)

102. M. Zhong, N. Goddard, C. Sutton, Latent Bayesian melding for integrating individual and population models, in *Proceedings of Advances in Neural Information Processing Systems*, Montréal, 7–12 Dec. 2015, pp. 3618–3626

103. I. Sutskever, J. Martens, G. Dahl, G. Hinton, On the importance of initialization and momentum in deep learning, in *Proceedings of the 30th International Conference on Machine Learning (ICML)*, Atlanta, 16–21 Jun. 2013, pp. 2176–2184

104. Theano Development Team, Theano: a Python framework for fast computation of mathematical expressions, *arXiv e-prints*, vol. abs/1605.02688, May 2016

105. F. Liebgott, B. Yang, Active learning with cross-dataset validation in event-based non-intrusive load monitoring, in *2017 25th European Signal Processing Conference (EUSIPCO)* (IEEE, Piscataway, 2017), pp. 296–300

106. J. Alcalá, J. Ureña, Á. Hernández, D. Gualda, Event-based energy disaggregation algorithm for activity monitoring from a single-point sensor. IEEE Trans. Instrum. Meas. **66**(10), 2615–2626, (2017)

107. M. Azaza, F. Wallin, Finite state machine household's appliances models for non-intrusive energy estimation. Energy Procedia **105**, 2157–2162 (2017)

108. I. Sutskever, J. Martens, G. Dahl, G. Hinton, On the importance of initialization and momentum in deep learning, in *Proceedings of the 30th International Conference on Machine Learning (ICML)*, Atlanta, 16–21 Jun. 2013, pp. 2176–2184

Index

© The Author(s), under exclusive license to Springer Nature Switzerland AG 2020   133
R. Bonfigli, S. Squartini, *Machine Learning Approaches*
to Non-Intrusive Load Monitoring, SpringerBriefs in Energy,
https://doi.org/10.1007/978-3-030-30782-0

Printed in the United States
By Bookmasters